James Long, John Benson

Cheese and Cheese-Making Butter and Milk

With Special Reference to Continental Fancy Cheeses

James Long, John Benson

Cheese and Cheese-Making Butter and Milk
With Special Reference to Continental Fancy Cheeses

ISBN/EAN: 9783743401426

Manufactured in Europe, USA, Canada, Australia, Japa

Cover: Foto ©Lupo / pixelio.de

Manufactured and distributed by brebook publishing software (www.brebook.com)

James Long, John Benson

Cheese and Cheese-Making Butter and Milk

Cheese and Cheese-making

BUTTER AND MILK

WITH SPECIAL REFERENCE TO CONTINENTAL
FANCY CHEESES

BY

JAMES LONG
JOHN BENSON

LONDON: CHAPMAN AND HALL, L<small>D</small>.
1896

[All rights reserved]

RICHARD CLAY & SONS, LIMITED,
LONDON & BUNGAY.

CONTENTS

CHAP.		PAGE
I.	THE PRINCIPLES OF CHEESE-MAKING ... By James Long	1
II.	THE TRADE IN FOREIGN CHEESE ... By James Long	12
III.	SOFT CHEESE MANUFACTURE ... By James Long	23
IV.	GORGONZOLA, AND THE VARIETIES OF BLUE OR MOULDED CHEESE ... By James Long	37
V.	OTHER VARIETIES OF FANCY CHEESE ADAPTED FOR MANUFACTURE IN ENGLAND ... ... By James Long	52
VI.	ON THE BEST METHODS OF MANUFACTURING CHEDDAR CHEESE ... ... By John Benson	66
VII.	ON THE BEST METHODS OF MANUFACTURING STILTON CHEESE ... ... By John Benson	80

CONTENTS

CHAP.		PAGE
VIII.	ON THE BEST METHODS OF MANUFACTURING CHESHIRE CHEESE By John Benson	94
IX.	ON THE BEST METHODS OF MANUFACTURING WENSLEYDALE CHEESE By John Benson	104
X.	THE MILK INDUSTRY By James Long	115
XI.	THE PRINCIPLES OF BUTTER-MAKING By James Long	127
XII.	CREAMERIES AND FACTORIES By James Long	138

CHEESE AND CHEESE-MAKING

CHAPTER I

THE PRINCIPLES OF CHEESE-MAKING

PROFESSOR HENRY, of the Wisconsin Agricultural College, recently stated that the loss of the American cheese trade with Great Britain was owing to the fact that his countrymen did not make the best article, and that in many cases imitation cheese was produced for the sake of a possible temporary profit, but to the ultimate loss of all concerned. Whatever may be the immediate gain effected by the addition of foreign fat to milk, or by the removal of a portion of the cream it contains, the permanent value of the cheese industry to the producer is maintained only by the manufacture of the best, and of its production in the largest possible quantity. To obtain both quantity and quality necessitates a close study of the subject and a recognition of the principles which underlie the practice of cheese-making. To obtain quantity of cheese it is essential to have rich milk. We are told by those who oppose the institution of a standard in this country that the solids present in milk do not exceed $11\frac{1}{2}$ to 12 per cent., but the cheese-maker who produces or buys milk of this quality will not find his returns very satisfactory. The value of rich milk to the cheese-maker is two-fold. In the first place, cheese is chiefly composed of the fat and casein of the milk—its two most important solids — and water; therefore, the more fat milk contains — and this is by far the most important constituent —the more cheese we produce per gallon, for three reasons: first, because the fat itself adds to the weight of the cheese; next, because with the increase of fat there is an increase of casein, which follows in an almost constant ratio; and last, it is a fact worth knowing that cheese produced from rich milk, *i.e.* milk containing a high percentage of fat, retains more water, and consequently weight is obtained from this source also. Every good cheese is mellow in its texture, and to some extent this mellowness depends upon the proportion of fat the cheese contains. Recognizing these facts, we come to the first principle which it is essential to remember, that in order to produce rich milk the cattle must be well selected, for quality depends rather upon breed than upon food. Nor is it entirely necessary to go to the Channel Islands for rich milkers. There are milkers of a very high order, as regards both quality and quantity, to be found in every British breed, particularly among Shorthorns and Devons.

It is, therefore, by selection and by testing the milk of cows retained in the herd, and excluding those which produce poor milk, that quality is maintained. Although, as we have remarked, breed has more influence than food upon quality, yet the production of fat in milk depends largely upon good feeding, inasmuch as good feeding improves the yield—although it may not increase the percentage of solids—and consequently it increases the fat. Thus we get to the soil, and it is usually found that in those districts where the most luxurious crops are grown—grass in particular, for it is the commonest food of cows—the cattle are best, and the milk they produce most abundant. Soil, however, has another influence which it is essential to mention. As we shall show, acidity plays an important part in the process of cheese-manufacture. But acidity is to some extent controlled by the alkaline properties which are present in milk, and as a proportion of these properties depends to a large extent upon the soil from which they are obtained, so does the soil indirectly influence the quality of the cheese, unless, by the exercise of the highest skill, sufficient allowance is made and the acidity controlled. Similarly, water exercises an influence when it contains an abnormal quantity of lime, and it is next to impossible to produce fine-flavoured cheese where such weeds as garlic are common on the pasture. The dairy, too, must be constructed with the object of providing perfect ventilation, the maintenance of an even temperature, and the exclusion of every possible means of conveying a taint to the milk.

Upon the first part of the process of manufacture in the dairy—that of the coagulation of the milk—a great deal depends. The period of the formation of the curd varies in accordance with the variety of the cheese produced. In the manufacture of soft cheese it is prolonged, sometimes for a considerable period; in the manufacture of pressed cheese it is usually short. The period of coagulation is influenced by the quality of the milk, the condition at the time the rennet is added, its temperature, and the strength and quantity of the rennet employed. The curd produced in a short time is elastic and comparatively firm; that produced after a prolonged period of coagulation is tender, it will scarcely bear cutting, and it parts with its fat, which is carried off in the whey unless it is very carefully handled. Thus it will be recognized that mellowness in cheese is obtained in different ways, but without sufficient moisture we can have no mellowness. Hence, if too large a quantity of rennet is used, if too much acidity is developed, or if the temperature is raised too high, the whey may be so rapidly and so completely expelled, that an insufficient amount of water will remain, either for the purpose of producing the necessary mellow condition, or even of ripening the cheese. In the manufacture of pressed cheese the whey is expelled by cutting the curd—and the finer it is cut the larger the surface exposed for its removal,—by heating to a high temperature, by the development of acidity—which causes the curd to contract—and by pressing. In the manufacture of soft cheese, however, the curd is not cut, except in such large slices as are essential for its removal into the moulds; but the whey drains off slowly by gravitation, and subsequently more is lost by evaporation. The cheese is soft because it retains more water than pressed cheese, while its flavour is largely influenced by the fact that it retains more sugar—the sugar being in solution in the whey—and because, in consequence, more acid (which is produced from the sugar) is developed. A tender curd, then, such as is generally used in soft cheese-making, is obtained by setting the milk at a low temperature and by the employment of a small quantity of rennet. In this way coagulation will be delayed. It is also essential that the milk used should be sweet, for if, as in pressed cheese-making, a portion of the milk used has been allowed to stand for a number of hours, acidity will have commenced to develop, which hastens coagulation, and will in time actually produce it.

The reason why curd which has been cut fine in the manufacture of large pressed cheese is left in the whey and heated, is that unless this were done it would not be sufficiently acid, for the curd when drawn from the whey is tough and dry as compared with the curd used in the manufacture of soft cheese. Unless this process were carried out the whey would not be expelled, and the cheese would not acquire its mellowness of texture or its fine nutty flavour. In soft cheese-making the curd is placed in small moulds; small cheeses are, indeed, essential, otherwise the whey would be unable to find its way to the surface; but unless the temperature is sufficiently high, it even then refuses to move, and for this reason soft cheese-making is conducted at specific temperatures which are applied to each variety of cheese. Theoretically, the time of coagulation is in inverse ratio to the quantity of rennet employed, but in practice this axiom is not entirely borne out, although the reasons do not detract from its truth. The same conditions do not, for example, apply to large quantities of milk, or to entirely fresh milk, which apply to small quantities or to milk which has been practically ripened by exposure. Thus, in the manufacture of small cheeses small quantities of milk are employed, and this milk parts with its heat more rapidly than is the case with a large volume. Again, when acid is developed slightly in milk, less rennet is required, and a milk rich in fat does not produce the same result with the same quantity of rennet as a milk poor in fat. It is important, therefore, in cheese-making to understand the quality of the milk employed, and, where it has been exposed for any number of hours, to ascertain the quantity of acid which it contains. Where small quantities of milk are set for curd, wooden vessels should be used, as wood is a non-conductor of heat; lids should be employed, and the whole covered with a blanket or any other non-conducting material.

We have referred to the nature of the solid matter of milk. The cheese-maker should early learn to understand that only a portion of these solids find their way into the cheese, the bulk of the sugar of milk, which forms a large proportion of the total solid matter, remaining in the whey, together with portions of the mineral matter, the casein, the albumin, and the fat. Almost the whole of the casein is, however, extracted in cheese-making, this being coagulated by rennet or by acid, whereas the albumin passes into the whey in almost all varieties of curd which are not submitted during manufacture to a high temperature, as it is coagulated only by heat. There is, however, a material which has been described by chemists as albumose, which always passes into the whey, not being coagulated either by heat, rennet, or acid. In accordance with the very extensive results obtained at the New York State Experiment Station, which we have had the advantage of inspecting, the average percentage of solids lost in cheese-making, *i.e.* by passing into the whey, amounts to 6·20, while the percentage of solids recovered from the milk, *i.e.* retained in the cheese, amounts to 6·30. The actual figures—from my *The Elements of Dairy Farming*—may, however, be quoted, as they are of considerable value—

Milk-Constituents lost in Cheese-making.

	Lost in Whey for 100 lbs. Milk.		
	Least.	Greatest.	Average.
Water	82·53	84·61	83·70
Total Solids	6·09	6·39	6·20
Fat	0·20	0·36	0·25
Nitrogen Compounds	0·68	0·76	0·73
Sugar, Ash, &c.	5·06	5·44	5·22

Milk-Constituents recovered in Cheese-making.

	Retained in Cheese for 100 lbs. Milk.		
	Least.	Greatest.	Average.
Water	3·10	4·08	3·68
Total Solids	5·95	6·72	6·32
Fat	3·19	3·63	3·41
Nitrogen Compounds	2·21	2·51	2·34

The term "nitrogen compounds" indicates casein and albumin. The largest proportion of solids which passed into the whey was in the months of August and September. The smallest proportion of fat lost in the whey was in June and July, whilst the smallest proportion of casein and albumin lost was in the months of July and August. Upon the basis of the work carried on at forty-eight cheese-factories, it was ascertained that 50·6 per cent. of the total solids of milk were recovered, including 90·98 of the fat and 75·71 per cent. of the casein and albumin. It has been supposed that a larger proportion of fat is lost when the milk is rich than when it is poor or of but moderate quality. But this is not the case, and the following table will show that the percentage of fat lost when the milk is rich is positively lower than when it is of lower quality; also that the percentage of cheese made is enormously increased as the milk increases in quality.

Group	Fat Percentage.	Lbs. Fat lost in Whey per 100 lbs. Milk.	Per Cent. Fat of Milk lost in Whey.	Lbs. Cheese made per 100 lbs. Milk.
1	3 to 3·5	·32	9·55	9·14
2	3·5 to 4	·32	8·33	10·04
3	4 to 4·6	·32	7·70	11·34
4	4·5 to 5	·28	5·90	12·85
5	5 to 5·25	·31	6·00	13·62

CHAPTER II

THE TRADE IN FOREIGN CHEESE

It is unnecessary to remark that our imports of cheese are very large; in a recent year, in accordance with a calculation which we made, no less than 5·7 lbs. of imported cheese were consumed per head of our population, as against 7·9 lbs. of home-made cheese, and the value of the cheese consumed in the same year per head of the people amounted to 6s., of which 2s. 4½d. went to the exporter. In 1892 we estimated the value of the cheese consumed in this country at eleven and three-quarter millions sterling, the home-produced article being valued at between six and a half and seven millions. The imports, however, have tended to increase, and if we take the month preceding that in which we write (1895) we find that the imports have reached 125,000 cwt., as against 71,000 cwt. in the same month of the previous year. Taking the average quality of milk, this import of cheese for a single month represents fourteen million gallons, or the produce of 35,000 cows, giving an average yield of 400 gallons each per annum. A simple calculation, based upon the average number of cows kept in any one district, will show how many of our farmers are displaced by the energy of the foreign producer, and the low prices he is willing to take.

The variety of cheese which is imported in the largest quantity into this country is made upon the Cheddar principle, although it comes from Canada, from Australasia, and from the United States, in each of which countries there is practically no rent to pay on the great majority of farms, while in very numerous instances the labour is performed by the occupiers themselves. Thus it is that we are under-sold, in spite of the cost of freight across the ocean. Next to Cheddar come the Dutch varieties, Edam, or round, and the Gouda, or flat Dutch. We have had the advantage of inspecting numerous farms in Holland, and of seeing the cheese manufactured, and we are in a position to understand how easy it is for the thrifty and industrious Netherlander to supply the British market, although he does so, to a large extent, with cheese of inferior quality.

Vast numbers of Dutch farmers are small owners, and live in the most frugal manner. Their cattle are deep milkers, and they feed upon extensive and luxuriant pastures, which are admirably managed, while the buildings forming the homestead are usually under one roof with the house proper, and are simplicity itself. It is not surprising, therefore, that Dutch cheese is sold at a low price. The Gouda variety is not unlike Cheddar when it is well manufactured, but in the majority of instances both Gouda and Edam are of second quality, whether it be as regards flavour or texture. Gorgonzola probably takes third place. This cheese, largely manufactured in Italy, is produced by very small as well as by larger owners of cows, who obtain their curd in a manner which is not altogether perfect, especially as regards cleanliness, and who work upon a system, if such it may be called, which is extremely crude and incomplete, although in the Italian schools a well-defined and perfect system is seriously taught by men of considerable attainments, as we have had opportunities of recognizing. As a rule, the Italian farmer does not complete the process of curing, and this applies equally to the large and costly Parmesan, which is manufactured so extensively in Emilia and Parma. There is a class of middle-men who are capitalists and who possess admirably arranged ripening cellars and caves, and these persons buy the white cheese—indeed it is often nothing more than green curd—the curing of which they complete. Among varieties of a still more tasty character, we have the Roquefort, produced from sheep's milk, although cows' milk is to some extent taking its place; Camembert, Brie, Bondon, Neufchâtel, and Port du Salut, all of which hail from France, the last-named being a partially pressed cheese, whilst the others are entirely unpressed and belong to the refined soft varieties.

At the market price of cheese, which has been very low for some time, the English farmer who makes a really good article probably obtains 5*d.* per gallon for his milk, net. There are, however, large numbers of makers who obtain less and who never make first-class cheese: there are some who obtain more and who have a reputation for a first-class article. In the Colonies and America it is probable that makers as a body do not receive more than 3*d.* per gallon for their milk, net. If, therefore, we take an average cow of moderate pretensions, giving 400 gallons of milk per annum (and it is an undoubted fact that the majority of the cows in the country do not exceed this modest quantity), we shall find that the returns per cow, taking 5*d.* as the basis, would amount to £8 6*s.* 8*d.*, while the returns in the Cheddar-producing countries abroad would only amount to £5. In a 40-cow dairy, therefore, the gross returns in England would amount to £366 per annum, and in the other countries referred to, to £200. The question now arises, whether this difference represents the extra cost of rent, taxes, and labour: whether, in fact, the farmer is better off in this country with the higher receipts, or in other countries with the lower receipts. We venture to think that the British farmer holds the superior position, and that it is better worth his while to pay a good rent for good land and an excellent equipment under a good landlord in England than to pay no rent at all—and we are speaking only of cheese-making—either on the prairie of America or in the Australian bush. We are quite aware of the fact that the figures we have taken do not absolutely represent the exact state of affairs in either country, inasmuch as cheese is not made throughout the entire season, but they are sufficient for our purpose, for in both countries farmers obtain somewhat higher receipts in the winter, either by the sale of milk in England, or by the manufacture of butter in America and the Colonies. Further, the cheese-making farmer adds to his returns by the production of pork, in the manufacture of which he daily employs the whey from the cheese.

The Dutch farmer does very little better than the Colonial farmer. As a small owner of land, he has no rent to pay, and as the labour upon his farm is confined to the management of the cows and a few pigs and the production of cheese, in which the wife of the farmer assists materially, there is little out of pocket paid in the year. The Italian farmers are not so fortunate as the Dutch; they are extremely poor, and the bulk of the profit of the cheese industry, which is very extensive, finds its way into the hands of the curers and middle-men. In France, however, at all events so far as the leading varieties are concerned, the farmers do much better, and in the past they have obtained golden success in the production of their finest cheeses, hundreds of men having bought the farms they occupy out of the profits they have made. It has been no uncommon thing, and it is not uncommon to-day, to find French cheese-makers realizing from 10$d.$ to 1$s.$ a gallon for all the milk they produce, through the medium of cheese. As we have urged for years, there are many varieties, some of which are well known in this country, which would have by this time enabled scores of English farmers to have followed their example. But, in spite of agricultural depression, in spite of the means of education which exist, and of the fact that we have introduced into this country the system of manufacture of a number of these varieties, systems which have been taught for some years now, we are not acquainted with a single practical farmer who has attempted to build up a business in any one variety, although there is an important market at his very door.

We have referred to a number of the varieties of cheese which are imported. Naturally, Cheddar stands at the head of the list as a British cheese. A pound of Cheddar is usually represented by about 10 lbs.—or a gallon—of milk; but the quantity of cheese made from a given quantity of milk depends upon the quality of the milk, and this varies both with the cow and with the month of the year. In the Somerset Experiments and the New York State Experiments at forty-eight factories, the following quantities of milk, in pounds, were required in the various months named to produce each pound of cheese—

	April	May	June	July	Aug.	Sept.	Oct.	Avge.
Somerset Expts.	12·4	11·8	11·5	11·1	10·9	10·2	9·7	11·01
New York Expts.	10·71	9·98	9·95	10·07	9·58	8·95	8·43	9·76

Thus we see that in Somerset, our great Cheddar county, the milk was richest in October, the month in which it was also richest in New York; but while it took considerably more than a gallon, on the average, to produce a pound of cheese in Somerset, it took less than a gallon in America; and in five sets of experiments carried out upon an enormous scale in the States, the milk was always richer than in the experiments in Somerset, which were carried out upon actual cheese-making farms. As regards Cheshire cheese, which comes next to Cheddar in this country, we have not the same exact data; no work upon the same extensive and well-considered scale having been carried out in the successful county of Chester. These varieties are pressed cheeses, and in the same category come the Derby, the Gloucester, and the Leicester cheeses, all of which are but variations of the great Cheddar type, having nothing really typical or characteristic about them when considered apart from their prototype. The unpressed firm cheeses made in this country are known as Stilton, Wensleydale, and Cotherstone, all of which are mellow and ripened by the aid of the blue mould which grows in veins within them. In making these varieties, slightly more milk is required to produce a pound of ripened cheese than is the case with Cheddar or Cheshire, and consequently the value is higher; but, owing to the extension of the system of dairy teaching, the two first-named of these varieties have been manufactured of late upon a much larger scale; so much so in the last year that if production is further extended, the new makers will have reason to regret their entrance upon the industry. They will find at the end of the season, when their harvest should arrive, that they have no market at any price; and I, therefore, venture to caution milk producers against entering carelessly upon an industry which is now overdone. Far wiser would it be to commence the manufacture of the Swiss Gruyère, the Italian Parmesan, or the French Brie, Camembert, or Port du Salut, for each of which the market is still supplied by foreign producers. Broadly speaking, the cost of producing Cheddar or Cheshire, Derby or Leicester, Dutch or Gruyère, all of which are pressed cheeses, is similar in amount; but immediately we handle the soft cheeses we reduce the cost of the milk required and increase the cost of labour. Abroad, old women are largely employed in the work, and are paid very small wages, these persons assisting the female members of the farmer's family. A Camembert cheese sells readily for 6*d*., and weighs about 11 ozs. A Brie, weighing 1½ lbs., or a little more, sells for 1*s*. 6*d*., also by retail. The quantity of milk required to make a Brie varies from two to two and a half gallons, and it may generally be taken as a standard that half a gallon of milk of a little more than average quality will produce about 14 ozs. of white or unripened salable cheese, or 12 ozs. of ripened cheese, these figures being liable to increase or decrease in accordance with the quality of the milk. There is a sale in London for Camembert and Port du Salut as well as for Bondon, Neufchâtel, and Gervais, all of which are very small cheeses, weighing a few ounces only, the first two being produced from new milk alone, and the last-named from a mixture of new milk and cream. The possibility of success depends upon the maker, for the London merchant is amenable to reason, and will buy in the English market if he can obtain a satisfactory article at a price which is at least not in excess of that charged by the Frenchman.

CHAPTER III

SOFT CHEESE MANUFACTURE

THERE is no doubt that the manufacture of soft cheese is the most profitable branch of dairy farming in France. We have for many years paid much attention to this subject, in the hope that the system might be established in this country; but, chiefly, perhaps, from want of knowledge of the system of manufacture, and to some extent from want of enterprise, our dairy farmers still allow the French to supply our markets, hesitating to take up a class of work which careful investigation would show them to be extremely profitable. The following remarks are not based upon theory; they are the result of a considerable amount of labour devoted to the study of the processes of manufacture of the leading varieties of soft cheese made in France. We were led to investigate the subject from the fact that no information was obtainable, and in spite of considerable help from personal friends in France we found great difficulty in arriving at really correct methods, while success was only achieved by continual experiment and practice.

BRIE.—In an article in the Journal of the Royal Agricultural Society, speaking of the Brie cheese, I pointed out that in five parishes in the Brie district alone six million cheeses were made annually. Assuming that each cheese weighed, upon the average, 4 lbs., this quantity represented the yield of 25,500 cows, assuming each cow to produce 450 gallons of milk per annum. Reference to the agricultural returns will show that in a large number of our English counties the cows kept do not reach this number. It has been urged that if every dairy farmer took up the manufacture of a particular kind of soft cheese the market would rapidly be overdone; but it is beside the mark to suggest what never has taken place and never will take place in connection with any industry, especially in this country, where farmers are proverbially careful in the extreme.

The prices realized for Brie in Paris are often considerable, sometimes reaching a shilling a pound. The Parisians are large cheese-eaters, consuming about 12 lbs. per head of the population per annum; and the money annually spent in the wholesale markets of Paris in this one variety of cheese alone is estimated at about four million francs. The Brie is a large, round, flat cheese, varying from three-quarters of an inch to an inch in thickness, and from 8 to 12 inches in diameter; but in a market like that of London, where the consumption is not large, chiefly, perhaps, because of the difficulty of placing the cheese before the public in prime condition, it is seldom offered in more than one size. In my own practice (for experimental work was followed by systematic manufacture) 10 lbs. of rich milk or $12\frac{1}{2}$ lbs. of ordinary milk were required to make a cheese which sold at 1s. 6d. The milk must not be skimmed, as the creamy character of the cheese is by this process very much diminished, as well as the mildness of its flavour.

The plant required in the manufacture of soft cheese is neither considerable nor expensive.

The draining-table should be made with a slight fall to the front, on the edge of which should be a narrow channel to carry off the whey; wooden tables are usually covered with metal, but slate or brick-built stands faced with cement are still better. In either case the whey is enabled to run by gravitation into the channel, and is carried by the same force into a receptacle made for the purpose. The floor of the dairy should be of smooth hard cement laid on concrete, and the walls either of glazed bricks or smooth-faced Parian cement kept washed with lime. The utensils necessary are round wooden tubs with lids, stools on which to stand them—preferably with rollers on the legs—a large metal skimmer without perforations, a thermometer, a rennet measure, moulds made of tinned iron the exact diameter of the cheese to be made, boards made of seasoned wood so that they will not shrink, and sufficiently large to place the cheeses upon, mats made either of rush or fine rye-straw and large enough to cover the moulds, a salt-dredger, and some round osier plaques or plates, called by the French *clayettes*. The plate is intended for the cheese to rest upon instead of a plain board, so that air may penetrate beneath it. The mould is in two pieces, the bottom having a rim into which the upper portion fits. The object of these two pieces is that the cheese may be conveniently turned, as we shall presently see.

In the process of manufacture, the milk is strained into a ten-gallon tub, wood being used to prevent loss of heat, and the rennet added at a temperature of from 82° to 86° F. A little practice will show the manufacturer which temperature suits his milk best, and which to adopt at different seasons of the year. The curd should be fit to remove into the moulds in four hours, the apartment in which the work is performed being kept at from 60° to 62° F. Great care must be exercised to set exactly the quantity of milk required for the manufacture of a given number of cheeses, and, as far as possible, each mould should be filled equally. Before moulding, the boards must be placed upon the draining-table, a dry, clean mat being laid on each, with the moulds on the top. The curd, which must be elastic, not sticking to the finger or the thermometer when inserted, is removed in large thin slices into the moulds. If the slices are thick the whey escapes with greater difficulty. When the moulds are filled the curd is left to drain, and in three to four hours, perhaps more in colder weather, the whey will have escaped and the curd have sunk into the lower portion of the mould. In this case the upper portion is removed, a mat is placed over the lower portion, followed by a board, the whole is rapidly inverted, the bottom mat and board removed, and subsequently cleansed, when the bottom of the cheese will be seen to be marked by the straws. On the following morning the same operation will take place again, so that the cheese will be marked on each side; but with this turning the new mat is placed so that the marks will be crossed, causing a number of little points to appear on the surface of the cheese, instead of lines. These points will subsequently be covered with mould. In a few hours the last turning takes place, and again in from four to six hours the curd will be sufficiently firm to stand alone; the mould will then be removed and the cheese fit to salt, this being done with extremely fine salt distributed by a dredger. Unless every portion of the crust receives salt the mould will not appear. Salting on the second side occurs some hours after the first salting: the cheese is then removed on its mat to a *clayette* and taken to the drying-room. . Here it stays for a few days, being systematically turned until it is covered with white mould. In some cases it may stay in this apartment: in others a third room will be essential for the development of the blue mould, which gradually appears until the whole of the cheese is covered, so that at the end of from three to four weeks it is salable. In France, however, consumers of Brie prefer it in an advanced state of ripeness, and the blue cheese is therefore taken to an underground cave until it becomes so creamy that upon the breaking of the crust it runs, and in this condition it realizes a higher price. I venture to think, however, that the English taste would prefer the blue cheese, which is milder and more substantial. No Brie is thoroughly ripe until the white and somewhat solid curd has become yellowish and creamy throughout. Ripening proceeds from the outside, and on cutting any soft cheese of this character while this process is going on, it will be seen, if the ripening is not complete, that while beneath the crust the cheese is creamy, in the centre it is still solid and to some extent insoluble. It has been pointed out by Duclaux, a French chemist of considerable eminence who has studied this question perhaps more than any other investigator, that the moulds which grow upon Brie and similar cheeses practically remove the acid present through the medium of what we may crudely term their roots, or mycelium, and that until this acid is removed the bacteria which are responsible for the ripening process are unable to complete their work.

CAMEMBERT.—Several years ago, I had the opportunity of inspecting a number of the most important Camembert dairies in the north of France, having already a close acquaintance with the system of manufacture. In one of these dairies—that of M. Roussel—1800 cheeses were made daily from 800 gallons of milk, the produce of 400 cows. I estimated at the time that if M. Roussel produced Camembert during only five months of the year he would turn out 107 tons of cheese, which at that time was realizing a somewhat extravagant price. It is therefore not surprising that the Camembert makers were able to save money and to buy the farms they occupied. From investigations made in the county of Calvados, in which Camembert is chiefly made, I learned that there were large numbers of farmers who each made from 10,000 to 160,000 cheeses per annum. There were 50 farmers manufacturing more than 25,000 per annum, and large numbers making smaller quantities. From the station of Lisieux 655,000 kilogrammes were dispatched; and from the village station of Mesnilmauger 12,500 cases containing 62,000 dozen. In some other counties the manufacture was also considerable, but now it is possible that it is doubled. Certain it is that Camembert is much more largely consumed, and that the bulk of the cheese which arrives in this country is produced from milk which has been partially deprived of its cream. Camembert was invented during the Revolution of 1791 by the ancestress of M. Cyrille Paynel, a large maker in Calvados, recently dead, whose acquaintance I made on my first visit to the district. It is well known in every part of

England, and would be certain to sell in much larger numbers than at present if its production were taken up as an industry. A gallon of rich milk produces about $2\frac{1}{4}$ cheeses, so that a cow yielding 600 gallons would make 1350 cheeses, which, at $4\frac{1}{2}d.$ each—which I believe to be the wholesale price of average cheese—would realize £25 6s. 6d. without the whey. The manufacture of Camembert, in a word, enables the producer to realize from 10d. to 1s. per gallon for his milk during the summer season, when Cheddar realizes only 5d. to 6d. a gallon (slightly more or less according to its quality), and butter about 4d.

The following is a description of the system adopted in the manufacture of the cheeses made in my own dairy, which gained the £10 prize at the Royal Agricultural Show at Newcastle, and the silver medal at the London Dairy Show. Seventy-five pounds of milk was set in the morning, and a similar quantity in the evening, at a temperature of 80° F. The quantity of rennet added to each lot was $2\frac{1}{2}$ cubic centimetres. The curd was fit for removal into the moulds in $8\frac{1}{2}$ hours. The moulds are small, deep cylinders, the inside diameter being equal to the diameter of the cheese. They are perforated, and are placed close together on an inclined draining-table upon large mats. A hundred and fifty pounds of the milk used, which, by the bye, was of high quality, produced three dozen cheeses; the 36 moulds were, therefore, nearly filled with the curd of the morning. In the afternoon the curd had sunk more than half-way down the moulds, which were again filled to the brim with the curd of the evening. On the following day, the curd having become partially firm by drainage, each mould was inverted on fresh mats. This is a somewhat delicate operation, and skill is only acquired by practice. Turning continues until the cheeses are firm enough for the moulds to be removed. They were then salted alternately on each side and placed in batches upon clean mats, which were laid upon boards made for the purpose, and left upon shelves which were fixed above the draining-table. Here they were regularly turned until the white mould commenced to grow, when they were taken to the *séchoir* or drying-room. In this apartment they remained until the blue mould commenced to grow, when they were removed to a cave, which was excavated in the chalk. Here great attention had to be paid to ventilation, and to the hygrometric condition of the atmosphere, and until this was perfected it was impossible to obtain first-class cheese; but once the condition was acquired there was no further difficulty. With the continued growth of the mould, ripening is pursued; insoluble curd becomes soluble, the flavour is acquired, and the cheese becomes fit for market. In some cases it may be necessary to heat the milk up to 86°, while some makers in France do not remove the curd until four hours, and others remove it in two. Small quantities of milk are always renneted in preference to large quantities. Great care must be taken in preventing a damp atmosphere either in the drying or ripening rooms. During fine weather both rooms are well ventilated, cross draughts being arranged in the former apartment, but during wet weather draughts are excluded and the room is kept as dry as possible. With excessive humidity the white mould changes to black, a variety known as the *Aspergillus niger*, while the blue mould, which is responsible for so much work in the process of ripening, is the common *Penicillium glaucum*—the shape of the tiny filaments known as hyphae, which are responsible for the propagation of the spores of the mould, resembling a painter's brush, hence the Latin word *penicillium*. It is curious that these tiny fungoid plants should have so important an influence in the ripening of cheese. The blue mould is unquestionably the dominant fungus in the atmosphere of the dairy; it will not only grow luxuriantly at the temperature at which soft cheese ripens, but at a still lower temperature when it is provided with a suitable soil or feeding material. It has been assumed by some writers that it is essential to cultivate the moulds common to cheese; but this is not the case. It is common to every household, and its spores or seeds are so easily dispersed by the movement of the atmosphere that wherever such a material as cheese is placed it is certain to be attacked. The maker of soft cheese should, therefore, observe the recognized rules of cleanliness which apply to all dairies: lime, boiling-water, and the scrubbing-brush being used with absolute freedom, and without any fear of eradicating the fungus, the aid of which is so essential to success.

CHAPTER IV

GORGONZOLA, AND THE VARIETIES OF BLUE OR MOULDED CHEESE

IT is curious that the public should hold opinions with regard to the production of the various cheeses having blue or moulded veins within, which are entirely unwarranted by the facts. I refer to such varieties as the Gorgonzola of Italy and the Stilton of this country. It is supposed by some that Gorgonzola, for example, is the product of goats' milk, or of the milk of the goat blended with the milk of the cow; and by others that the blue mould is introduced by the insertion of metal skewers, which, by the way, are sometimes used, and used, too, for the purpose indicated, although the result is achieved in a very different manner from that supposed. The blue mould of cheese is the common *penicillium* which attacks bread and other materials common in the household. It is probable that it is abundant in every apartment of a house, and nowhere more so than in the dairy where cheese is made. If we regard the mould as a plant, and that plant as a weed, we shall better understand the principle which is followed in its extensive production by remembering that as the seeds of weeds are more prolific in the production of plant life when they fall upon fertile soil (such as the well-tilled and well-manured arable land of the farm) than when they fall upon the highway, so does the tiny plant which we call mould increase with great rapidity when it alights, as it were, from the atmosphere upon curd, which to it is a most fertile soil. It grows, elaborates its seeds or spores, which in their turn are shed abroad, falling upon similarly fertile soil, the curd of other cheeses, ultimately covering the portions in which they are permitted to grow.

GORGONZOLA.—Gorgonzola cheese is made from average cows' milk of the northern part of Italy, in which country I had the advantage of learning a great deal about the system. The cows' milk of Lombardy, to which reference is chiefly made, is not so rich as is generally supposed, but it is not absolutely essential that the milk intended for conversion into Gorgonzola or Stilton cheese should be specially rich in fat. To a very large extent this milk is produced by small owners of cows, who manufacture the cheese, but do not perfect or ripen it, selling it to merchants for this purpose, who in their turn finish the process in the cellars and caves which they own. Gorgonzola is a cheese which is produced from two curds, that is to say, from two lots of curd made at different times. When the two curds are put into the mould which gives form to the cheese, one is cold and stale and the other warm and fresh. For example, assuming the cheese to be moulded in the morning, the milk of the evening previous having been brought to a temperature varying from 80° to 85° F., and in some cases 90° F., the rennet is added. It is important, however, to make one or two remarks at this point. In dairies which are conducted upon defined principles the temperatures adopted are systematically arranged in accordance with the weather; but large numbers of small farmers who have no dairies worthy the name, add the rennet to the milk just as it comes from the cow, so that the temperature may vary from 90° up to 93°. Again, the rennet generally used in Italy is a filthy preparation which is, practically speaking, the macerated stomach of the calf, the actual animal matter itself. A portion of this material is placed in a piece of cloth and dipped with the left hand into the milk, the right hand the while squeezing it in order that the extract which exudes may be mixed with the milk, which is subsequently stirred. In Italy the curd, when fit for cutting or breaking, is gently broken with an instrument called a *paumarilo:* the operation lasting about a quarter of an hour. The whey is gradually expelled until the curd is fit to be hung up in a cloth on to a hook in the ceiling, and there left until the following morning. It is essential that the apartment in which it hangs should be at least 60°, but not more than 65° F. If higher, it may become too dry; if lower, too heavy, the whey not leaving it properly. Naturally, however, difficulties are met with by the small dairymen in the mountainous districts, especially those who are constantly moving with their herds of cattle, and therefore compelled to make the cheese wherever they may be; this system it is which accounts for so much inferior Gorgonzola.

The curd of the morning is in the first place treated in a similar manner to that adopted with the curd of the evening, but when broken every effort is made to obtain from it a large quantity of whey while it is still warm. A small quantity of acid forms in the evening's curd, but the curd of the morning should be perfectly sweet. The mould used in the manufacture of Gorgonzola is a curled piece of wood, preferably beech; but in some cases metal is being introduced in consequence of the fact that it can be more easily cleaned, not absorbing the whey, as is the case with wood. To one end of the mould a cord is attached, so that the cheese may be tightened or loosened as may be found desirable. When ready for moulding the curd is placed on the draining-table, which is fluted to carry off the whey, and the mould is placed on a rye-straw mat. Sometimes the mould is divided into two parts, the upper portion fitting into the rim at the head of the lower portion, and being removed when the curd sinks. Before filling, the mould is lined with a strainer cloth. In commencing, the bottom of the mould is covered with a thin layer of the warm morning's curd. Above this is placed a layer of the curd of the previous evening, followed again by another layer of warm curd, and so on until the top is reached, care being taken that the warm curd covers the entire surface of the cheese. The prime object, as I believe, of thus alternating the two different kinds of curd is that the mould is enabled to grow in the interstices which are formed, inasmuch as the warm and cold curds never unite in the same close, homogeneous manner as is the case where the curd is all made from one lot of milk, and is all of one temperature.

MOULD-RIPENING.—In different countries different methods are followed for the production of the mould. For example, in that part of France where the famous Roquefort cheese is produced from the milk of the ewe, the makers do not rely absolutely upon its natural growth, but they specially prepare a kind of bread, which is crumbled, and upon which mould is induced to grow, which it will easily do by exposure to a slightly warm, humid atmosphere. The mouldy crumbs which are thus produced are mixed with the curd, which is subsequently converted into cheese.

After the cheese has been formed it remains for drainage in an apartment at about 66° F. It is frequently turned, taken out of the mould, the cloth changed, and turned again. In Lombardy, where the cheese is sold in its new or green form, it is weighed at the time it is last taken out of the mould. It is then ready for removal to the salting-room, where it subsequently remains a few days at 68° F. The cheese will then be found covered with a fine growth of white fungus, which is an indication that it is ready for salting. The finest salt is used by the best manufacturers, although those who exercise little care use any salt which comes to hand. The surface of the cheese is entirely covered by gently sprinkling, the salt being subsequently rubbed into the crust with the hand. As a rule, this method of salting continues daily for a considerable period, from two to four weeks; but in some cases the upper portion of the cheese is salted at one time and the lower portion at another, that is, on the following day, so that the entire cheese is really salted from twelve to fifteen times. When this process has been completed, the texture of the cheese may be examined. If it is too close, it is possible that the fungus or blue mould will not grow with freedom. In this case the cheese is pierced with metal skewers, which admit the air, and with it oxygen, which the fungi require, for they are unable to grow in its absence. Should the texture, however, be sufficiently light and generous, nothing need be feared, as it will grow equally as well as in the Stilton, in which the texture is generally closer and mellower.

When Gorgonzola cheeses are taken to the cave to ripen—and some of the Italian caves which we have been enabled to see are very fine and well arranged—they are laid upon shelves covered with rye-straw and kept at a temperature of about 55° F. As with other cheeses, ripening can be hastened by a rise in the temperature, but the best cheese is that which is produced during the process of a longer time, and at a lower temperature. During the ripening process, which may take as long as from four to five months, or even more, different varieties of fungi grow upon the crust. The first to appear is a fungus of a dark colour, which is followed by a white mould, and subsequently by a red fungus, which is supposed to give colour to the cheese, although this colour is generally simulated by artificial means. The best Gorgonzola is of a very high type indeed, but it is seldom seen in this country.

STILTON.—The leading blue moulded cheese in this country is the famous Stilton, and the system adopted in its manufacture is not unlike that which is followed in Italy in the manufacture of Gorgonzola, or in France in the manufacture of Roquefort and several other varieties of a similar character. Stilton is the leading cheese of a class which in this country includes the Wensleydale and the Cotherstone, both of which when really perfect are varieties which it is difficult to beat; indeed, a perfect Wensleydale, with its mild flavour and mellow texture, is scarcely equalled by a perfect Gorgonzola, and I am not sure, although Stilton is made in my own dairy, that this more famous variety can at its best equal either of those named at their best. It is, however, fair to say that perfect cheeses of either kind are much less often seen than is the case with Stilton, in the production of which very considerable skill is now brought to bear, the industry being one in which there is keen competition, and which, in consequence, it is to be feared, will in the future yield lower prices to those who produce this cheese. There are different methods adopted in the production of Stilton, which it is proverbially supposed can only be manufactured with success in Leicestershire. This, of course, is fallacious; but there is a great deal in Leicestershire herbage, if not in Leicestershire cattle or climate. A method which will be found successful is that of setting the morning's milk at 85° F., and removing the curd in thin layers at the end of an hour into the draining-cloths which are laid upon a properly constructed draining-table. It should be observed, however, that in no case is it possible to lay down definite figures for all cases, whether they relate to temperature, time,

GORGONZOLA, AND MOULDED CHEESE 47

or the quantity of rennet used. The quality of the milk and the climate of the district have considerable influence, and these influences must be met by a slight deviation either in the temperature at which the milk is set or the quantity of rennet added, to say nothing of one or two subsequent details. The curd then is placed layer by layer into the drainers. Here, being warm, it gradually parts with its whey, and as it becomes firmer the corners of each cloth are tied loosely together, in order that the slight pressure thereby exerted may cause the whey to leave it still more effectually. These corners are from time to time tightened until the curd is fairly firm, and can be handled without breaking into pieces. When the temperature of the air is about 60° F. the curd may be generally left throughout the night, but when the temperature is below 60°, the curd had better be slung in a cloth from the ceiling, as suggested with regard to the Gorgonzola. In this way the curd parts with its whey more freely. On the following morning it may be removed, cut in cubes, and laid in an open shallow tin vessel to air. Airing is a somewhat indefinite term, but it may be mentioned that the object is to create or increase acidity in the curd. There can be no acidity without contact with oxygen, and as the air contains oxygen, so the curd is aired.

The morning milk is treated in a similar manner, and sometimes on the evening of the day on which this curd was produced it may be placed within the mould, but it depends upon its condition, for it must not be broken up for moulding until it is sufficiently firm and ripe, more particularly if the weather is cold, as in this case the cheese would swell and be utterly spoiled. On the second day, however, it is always possible to mould. The mould used is a cylinder slightly larger in diameter than a Stilton cheese itself. It is perforated with a number of rather large holes, through which a certain quantity of the whey exudes when the curd is within it. The mould is placed upon a cloth and is gently filled by the hands with the mixed curd of the two milkings. At this time the earlier curd is distinctly acid both in taste and smell, and also silky and mellow. Before mixing, both curds are broken into fine pieces with the fingers as gently as possible, and, after weighing, mixed with a fair proportion of salt. It is salted curd, therefore, of which the cheese is made, and in this particular, as well as in others, it differs from the Gorgonzola process. Both top and bottom of the cheese are carefully finished off so that the edges are cut clean and the surface level. In the course of three or four days, should the temperature be maintained at from 60° to 63° F., the cheese will be firm, and will have left the sides of the mould, which may be lifted from it, allowing it to stand alone. It is now bound with a calico binder somewhat tightly, and pinned top and bottom. This bandage is removed and a clean one put on every day until the somewhat wrinkled coat of the cheese has partially formed. It is then taken to the drying-room and subsequently to the ripening-room.

All cheeses of this character lose considerably in weight, in spite of the fact that they are not pressed, and yet they maintain a mellower, softer, creamier texture than cheeses which have been pressed. It is possible to hasten the process of ripening : first, by drying the cheese at a slightly higher temperature than is common, and next, by ripening it in an apartment kept at from 65° to 67° F., and pronouncedly humid. On the other hand, ripening may be delayed by the adoption of a lower temperature, which both prevents the mould from growing so freely, and the bacteria (which play an important part in the conversion of the insoluble curd into soluble cheese) from carrying out their work so rapidly.

New makers are apt to take up a variety of cheese, the producers of which are already numerous. The Italians are producing more and more Gorgonzola, while in England, Stilton, being the most fashionable of the blue moulded cheeses of this country, has had the ranks of its makers reinforced so much of late, that the price has fallen to such an extent that the industry will presently not be worth following. There is great room for the extension of the system adopted in Wensleydale, and it is certain that if this cheese were systematically produced, and if it were mild and mellow as the very finest of the samples are, it would be much more largely sold than is possible under present conditions, under which its sale is almost localized, and its existence practically unknown in many parts of the country, to say nothing of the other English-speaking countries of the world. The manufacture of all these varieties is taught at the British Dairy Institute, Reading, and we are in a position to know that the instruction is really worthy of the attention of those engaged in dairy work.

CHAPTER V

OTHER VARIETIES OF FANCY CHEESE ADAPTED FOR MANUFACTURE IN ENGLAND

THE term "fancy cheese" has usually been applied to varieties produced from cream or full milk, or a mixture of cream and milk, which are small in size by comparison with the large cheeses of all countries, and which are unpressed, or only partially pressed, in the course of manufacture. But the Americans have applied the term to some cheeses which are pressed and which really have no claim to it in any sense of the word. Sometimes a private maker, who has a considerable reputation as a prize-taker, and who is in consequence enabled to obtain high prices, is termed a maker of "fancy" cheese for the simple reason that his product is exceptionally excellent, and that it is obtainable only by those who are willing to pay the price for it.

It should be the duty of every maker to endeavour to produce fancy cheese in this sense, but there is no fear of the article being placed before the public in too large a quantity, as there are comparatively few makers who excel, the great majority producing cheese of second quality. Fancy cheese has not been produced in this country to any considerable extent. We have already named a few varieties; there are, however, others which are worthy of the consideration of the manufacturer. On the Continent, and more particularly in France and Italy, there are numbers of small cheeses of various types produced in different localities, each of which has its admirers who consume it in large quantities, and who pay the producer a relatively larger sum per pound than is obtained by the makers of the huge pressed cheeses of Great Britain, America, and the Australian Colonies. Let us refer to some of these varieties. We have already mentioned the famous Gruyère of Switzerland, the Parmesan of Italy, both of which are pressed cheeses of considerable size; we have also referred to the blue cheeses made in our own country, to the Gorgonzola of Italy, and the Roquefort of France, as well as to the two leading soft cheeses made by different sections of the French people, the Brie and the Camembert. These varieties may be supplemented by the Port du Salut, Pont l'Evêque, and Neufchâtel, the Gervais, Coulommiers, and Bondon, all of which are made in France.

PORT DU SALUT.—The Port du Salut has long been one of the most delicate and popular varieties made upon the Continent, but although there are numerous makers, those who produce the perfect article are extremely few in number. The system of manufacture has until recently been supposed to be the secret of the Trappist monks, a colony of whom are located at the Monastery of Bricquebec, in the Department of Manche. A few years ago I had the pleasure of accompanying to the north of France a party of our own countrymen who desired to see something of the dairy system pursued by the most successful among the Norman farmers. We were enabled to see a great deal in consequence of the kindness and liberality of several of the farmers and others with whom I was previously acquainted. But my application to the Monastery, although backed by an introduction from one of the highest officials in the French Agricultural Department, was met by the response that no outsider was ever allowed to see the process of manufacture pursued; that, in a word, the monks could not trust their own friends, who under the guise of curiosity had in previous years apparently taken advantage of the privilege extended to them to describe something of the system pursued, and thus to place other people in possession of a secret which is so jealously guarded. Secrets of this kind, however, are not long-lived, and it is impossible to prevent those who are acquainted with the principles of cheese-making from producing a variety of this character if they care to take the trouble to make a few thoughtful and well-arranged experiments for themselves. The Port du Salut cheese is not unlike a variety made in this country and known as the Caerphilly; it is circular in form, flat, about an inch in thickness, and partially pressed. The *pâte*, or flesh of the cheese, is extremely mellow or creamy, and yet homogeneous and firm in consistence, although there are a large number of holes throughout, which are characteristic of the variety, and which, in proportion to their size and number, are concurrent with its flavour. The milk is brought to a temperature of 86° F., and sufficient rennet is added to bring the curd in thirty minutes. The temperature is slightly varied with the season, as with almost every other variety of cheese, while the rennet used is in proportion to the quality of the milk. The curd, which is primarily deprived of a portion of its whey by gravitation, is subsequently enclosed in a mould which is lined with a strainer-cloth, and subjected to slight pressure. The press generally used is of a very simple character; a number of screws are placed side by side on a beam, several cheeses being pressed at the same time. The screws are really turned by hand, so that it will be seen in a moment how slight and simple the process is. Port du Salut, having been deprived of its superfluous water, is ripened at a temperature of 54° F. The object is to prevent it becoming dry, and to ensure that slow process of change which is brought about by bacteria, so that it will be soft, mellow, nutty, and yet mild in flavour. This variety is already sold in England, and it is appreciated in London, where it is growing in favour. It is one of the most delicious cheeses, and its character is such that if it became better known to the English people it would be more highly appreciated, and would obtain a considerable sale. I know of no variety which is more worthy of production, and those who take it in hand will not only find that it is easily made, but that it will return them a profit far in excess of anything which can be obtained by the manufacture of the pressed cheeses which are made in such large quantities.

PONT L'EVÊQUE.—Pont l'Evêque cheese is a variety with a great local reputation in the north of one of the most important dairy departments of France. It takes its name from a village not far from Havre and Lisieux, and is sold in considerable quantities in the fashionable watering-places of Trouville and Deauville. I was enabled to see the system pursued by the most famous maker, a highly intelligent farmer, upon his own farm near Pont l'Evêque. This cheese, although unpressed, is firmer in texture than either the Brie or the Camembert, owing to its being deprived of its whey with much greater rapidity. The cheese is either square or oblong, slightly less than an inch in thickness, and weighing from 14 to 17 ounces, for the size is not uniform; its crust is comparatively tough, and it may be kept for a considerable time with safety. Practically speaking, a gallon of milk will produce a good cheese, but as milk varies considerably in quality, it follows that very rich milk would produce a much larger cheese than poor milk. The milk is set at a temperature of 88° F., with sufficient rennet to bring the curd in fifteen minutes. A large rush or rye-straw mat is laid upon the draining-table. This mat may measure a yard in length by 26 to 30 inches in width, in accordance with the quantity of curd handled. When the curd is firm enough to remove, it is gently cut in cubes of large size, and with equal gentleness removed with a metal dish on to the mat, where it immediately commences to part with its whey. As the whey runs off, the curd toughens, the ends of the mat are drawn together, the slight pressure involved causes a still further loss of whey, and this goes on until the curd can be handled and placed in the metal moulds, which are made in accordance with the size the cheeses are intended to be. The newly-moulded cheese is then placed upon a small mat, and on the evening of the first day turned on to another mat. The result is that both sides of the cheese are free from fractures, the curd being homogeneous, and both are marked with the straws. It need hardly be added that where a large number of cheeses are made the mats are numerous and large, and provision is made for the moulds to stand side by side in order that space may be economized. Turning goes on from day to day until the metal mould is removed. Fungi then gradually appear on the outside of the cheese until it is ultimately covered with blue. This growth depends upon the temperature adopted: in the first stage of manufacture the temperature of the dairy is 63°; when the cheese is removed into the first ripening apartment it is kept at 58°, and when it is taken to the cave for slow ripening, it is kept at 56°. Here, again, the apartment should be slightly humid as well as cool, one reason being that it is essential to maintain the moist character of the cheese, and to prevent the evaporation which, if allowed to continue, would ensure its being dry, unpalatable, and unsalable.

GERVAIS.—The Gervais cheese is a delicate little luxury produced upon an enormous scale by several makers in France, two of whom are pre-eminent, M. Gervais and M. Pommel, both of Gournay. These makers produce millions in the course of a year. M. Gervais supplies Paris, sending up fabulous numbers every day; M. Pommel, I believe by private arrangement with his neighbour, supplies other markets, including that of London. I have paid a visit to both establishments, and was able to see a great deal that was interesting and instructive in the factory of M. Pommel. Gervais is a mixture of cream and milk; it is unnecessary to suggest what proportion should be used, inasmuch as every maker has his own idea, but one-third of average cream mixed with two-thirds of whole milk will produce a most palatable and luxurious cheese. The essence of this system is the low temperature at which the mixture is set, 65° F.

The rennet added is so small in quantity—it is also mixed with water—that coagulation is not complete for from eight to ten hours: indeed, one maker made a practice of delaying coagulation until twenty-four hours. The object after the removal of the curd is to extract the whey, and one of the simplest plans is to suspend it in a cloth or bag until it is sufficiently firm to be removed to the Gervais press. The somewhat firm curd is laid in a cloth, which is placed within a slatted wooden frame from six to nine inches in depth, and a heavy wooden block is then placed upon it: examination takes place from time to time until the curd is perfect in texture. It is then placed in batteries of little moulds which have been already lined with specially made unglazed paper—in order to envelop each cheese—on the outside of which the maker stamps his name and address. These cheeses are extremely profitable, and, partaking so much of the character of cream (with which the flavour of the cheese is combined), they are readily salable at a remunerative price.

BONDON.—Bondon cheese is largely made in the country districts around Rouen. It is produced entirely from milk, and is an important industry among the very small farmers and cottagers of that part of France. Once, upon a visit to a large farm in the district, I was taken to see the dairies of a number of the smaller occupiers, whose wives my conductor systematically but fraternally kissed, and who were really the makers. Bondon, like Gervais, is extremely small, and from seven to nine cheeses are made from one gallon of average milk. The milk is set at a low temperature, and the curd takes a long time in coagulation. It is removed when firm to a strainer-cloth which has been stretched by the four corners over a vessel somewhat resembling an ordinary washing-tub. Here it gradually parts with its whey, being occasionally and gently moved, when the curd forms a coat which prevents the passage of the whey through the cloth. At a certain stage it is removed into a clean cloth, which is folded over it, covered with a board, and gently pressed. The right consistence having been obtained, the little cheeses are moulded by hand in a most expert manner, the mould being a small copper cylinder some three inches in length by an inch and a half or thereabouts in diameter. I am bound to say that the process is difficult for an inexperienced maker, but like every other difficulty, it can be overcome by patience and practice. The cheeses are subsequently salted, and either sold at the end of a week in their fresh and white form, or kept in a cave until they have been covered with mould, when their flavour is enhanced and their value increased. They are sent in trays to the markets, the smaller makers sending weekly or fortnightly, and the larger makers nearly every day. In the manufacture of the Neüfchâtel, which resembles the Bondon in form, care is taken to prevent the curd being too close and homogeneous; the curd is drained without pressure, and in consequence of its lighter texture when moulded, the spores of the common blue fungus, *Penicillium glaucum*, are enabled to develop during the ripening process, so that the interior of the cheese is blue as a Stilton and is prized in consequence, realizing a higher figure in the market.

For some years several of these varieties have been sold in the London and other markets in considerable numbers, but these quantities do not represent what would be considered an extensive industry were they produced in this country. Coming from France, they realize prices which, in consequence of the cost of carriage, are, perhaps, a little more considerable than they need be. If, however, we remember that a cheese which can be made at the rate of seven or eight to the gallon of rich milk, as is the case with the Neufchâtel, realizes 3*d.*, it follows that the remuneration which the farmer obtains by producing a cheese of this character is very considerable as compared with the small prices which milk obtains in the open markets.

Lastly, a few words about the Coulommiers cheese, which is made in the Brie district. I believe this to be one of the most important and most delicious cheeses made on the Continent, and it was the first the manufacture of which I introduced into this country. The first lessons I received in the principles of its production were given me by a very famous maker, Madame Decauville, of Coulommiers, who produces an article of the very first quality. It resembles the Camembert in form, but is slightly smaller in diameter, and thicker. It is made upon the Brie principle, and may be sold new at the end of a week with great advantage, for in this state it is much appreciated by the people of England; but ripened, and sold at the end of six or seven weeks, it is infinitely more delicious, and will return from 11$d.$ to 1$s.$ per gallon for all the milk utilized in its production.

CHAPTER VI

ON THE BEST METHODS OF MANUFACTURING CHEDDAR CHEESE

THE making of a good Cheddar cheese depends largely on conditions which are conveniently summarized by the word "medium." A first-rate quality of Cheddar can be made in any district, provided that you have soil of medium quality, which will grow a short, sweet herbage. Soils resting on and derived from limestone rocks are ideal; yet any soil of fair body, growing herbage free from all coarse grasses, &c., and containing a small percentage of leguminous plants, is equally appropriate. The breed of cattle is of considerable importance, owing to the great variation in the nature and quality of the milk which they yield. Those yielding milks rich in fat, and with a great difference between the size of the largest and smallest fat globules, are not so suitable as those yielding a milk containing an average percentage of fat, with only a slight difference between the size of the fat globules. When a milk is rich in fat there is danger of loss during the making of the cheese. When the fat globules are nearly uniform in size, you are able to get a more perfect distribution of them throughout the cheese. The milk of different breeds varies in colour, some yielding a milk almost white, others one decidedly yellow. The nearer white the milk the better, if artificial colouring of the cheese is not going to be practised. A typical cheese-making milk is that of the Ayrshire breed.

The food which the cow receives influences the milk. The ideal food for producing a cheese-making milk is grass; and the addition of cake to the diet of a cow renders the milk more suitable for butter than for cheese-making. This is because prime Cheddars are made from a medium quality of milk rather than from an excessively rich one. Besides, the increase in the richness of milk from such feeding is largely that of the fat of the milk, and consequently no appreciable increase in the quantity of cheese is obtained; whereas if butter was made a corresponding increase in the butter yield would be got. Again, cheese made from the milk of cake-fed cows is liable to deleterious changes during manufacture. The drinking water of the cows should be free from all suspicion of contamination. Water from stagnant ponds, or the effluent water from sewage farms, renders cheese liable to become spongy. The surroundings of the cow must be clean. The chief cause of complaint against milk is probably due to contamination after it is drawn from the cow. Given a suitable district, breed of cow, food, water supply, and surroundings, the cheesemaker can depend on commencing with a first-class raw article, *i.e.* a milk of average quality, suitable colour, with uniformly sized fat globules, and free from contamination either in the form of injurious bacteria or acquired taints.

A Cheddar is a whole milk cheese, and consequently no fat is extracted from the milk which is intended for its making. The evening's milk is strained into the cheese-vat, and kept at 64° to 68° F. The temperature is varied according to the conditions of the weather and the keeping qualities of the milk. In the morning the cream is skimmed off, heated to 90° F., and returned to the vat through the strainer along with the morning's milk. By this plan we get thorough mixing of the cream off the evening's milk, with the mixed evening's and morning's milk. The milk is now allowed to ripen, if it is not already ripe enough.

RIPENING is essentially acidity development. There are two methods of attaining the desired result. (*a*) The old Cheddar method in which a certain amount of sour whey is added to the milk in the vat. This is an empirical plan which does not take into account the amount of acid already present in the milk, and also risks one day's contaminated whey tainting the rest of the season's make of cheese. (*b*) The more modern method, and that adopted by the Canadian makers, is to keep the milk at a certain temperature (90° to 95°) until the required acidity develops. This temperature is the one that is most favourable to the growth of the bacteria which produce the acid we desire to obtain.

TESTING FOR ACIDITY.—There are two methods by which to determine the ripeness or amount of acidity developed—(*a*) By means of rennet. Take 4 oz. of milk at the temperature at which it is intended to rennet the milk, and add 1 drachm of rennet; if the milk coagulates in 20 to 22 seconds it is ready for renneting. (*b*) By means of chemical re-agents. Take out 10 c.cs. of milk with a pipette, run into a white porcelain dish, and add three drops of phenol-phthalein solution (addition of an alkali to a solution of phenol-phthalein produces a pink coloration). From a burette allow to drop soda solution of such strength that 1 c.c. of it will neutralize 0·01 gramme of lactic acid. Whilst adding the soda solution, keep constantly stirring the milk in the dish, and on the appearance of the faintest tinge of pink which remains permanent, you know that the whole of the lactic acid in the milk is neutralized. If it requires 2 c.cs. of the soda solution for this purpose, we know that we have 0·2 per cent. of acid in the milk, which is about the correct amount for making Cheddar. The former of these methods is probably to be preferred, owing to its requiring materials which are always at hand, and similar materials to those you are going to use in the actual cheese-making. The ripening or development of acidity is done with the object of aiding the coagulating action of the rennet, to assist in expelling moisture from the curd, and to shorten the whole process of manufacture.

RENNETING.—Assuming that the correct amount of acidity is developed, and that the temperature of the milk is 82° to 85°, depending on the season of the year, the atmospheric conditions of the day, &c., we add a sufficient quantity of rennet to ensure coagulation in 45 to 60 minutes. Usually 4 to $4\frac{1}{2}$ oz. of Hansen's rennet extract to each 100 gallons of milk is sufficient. After thoroughly stirring the milk and rennet, cover the vat with a cloth, and leave the curd until firm enough for cutting. When the curd makes a clean break over a finger inserted under and along its surface, it is ready for cutting. If cut before it is firm enough, you get a white whey owing to loss of fat, and this will happen however carefully the cutting is performed. If, on the other hand, the curd is too firm, you require to use such force in cutting that you also get a white whey, owing to the injury done to the curd.

CUTTING.—In the old Cheddar system a large single-bladed knife was used. In the Canadian system American cutters are used. With the latter the curd is first cut with a vertical knife lengthwise and crosswise, then with a horizontal knife in the same manner. Clean the sides and bottom of the vat with the hands; cut again with two knives both ways, and allow to settle ten to fifteen minutes, the shorter period if the curd is hard, the longer if it is soft. The object of cutting is to facilitate the escape of the whey, and cutting into uniform-sized cubes aids in the securing of a good curd.

BREAKING.—After settling, stir the curd carefully with the shovel breaker or rake for fifteen to twenty minutes, until the curd is the size of peas, and thoroughly intermingled with the whey. Then commence the application of heat or scalding, which usually takes place some forty minutes from the time cutting commences.

SCALDING.—This is done to render the curd

METHODS OF MANUFACTURING CHEDDAR 73

firm, and to develop acidity. There are two methods of scalding—

(*a*) The old method in which the operation is performed in three stages. The process consists in drawing off a proportion of the whey, and after heating it to a certain temperature adding it slowly to the contents of the vat. This is repeated three times. The first time the whey is heated to 110°, the second to 120°, and the third to 130°. The temperature of the contents of the vat is raised the first time to 90°, the second to 95°, and the third to 100°. To ascertain the number of gallons of whey to draw off, multiply the number of gallons of milk at the commencement by the number of degrees it is intended to raise the contents of the vat at the first scald. This product, divided by the number of degrees of heat it is intended to raise the whey, gives the number of gallons of whey required ; *e.g.*—

Contents of vat, 100 gallons.

Temperature to which it is intended to raise the contents of the vat, 90°.

Temperature of whey before commencing heating, 85°.

Temperature to which it is intended to raise the whey, 110°.

Thus we have—

$$90° - 85° = 5° \times 100 \text{ gals.} = 500$$
$$100° - 85° = 25°$$
$$\left\{ \frac{500}{25} = 20 \text{ gals., amount of whey required.} \right.$$

The contents of the vat are stirred fifteen minutes after each scalding, but after the last scalding stir until the curd is sufficiently cooked.

(*b*) The more modern method (which requires a jacketed vat and steam) is to raise the temperature continuously at the rate of 1° in three minutes, until 100° is reached, and then keep it at 100° until the curd is sufficiently cooked. Scalding ought to be done more slowly if little acid is present in the curd, and more rapidly if the acid is well developed.

The curd is known to be scalded sufficiently when it is shotty, hard, sinks quickly, has an acid smell, and answers to the hot iron test. This last test is simple and gives constant results. It is performed by taking a small quantity of

METHODS OF MANUFACTURING CHEDDAR 75

curd, compressing it tightly in the hand, drying it on a cloth, and then applying it firmly to a bar of iron heated to black heat, and gently drawing it away. If acid enough, the curd attenuates to fine threads of $\frac{1}{4}$-inch length. If not acid enough, it will not so attenuate; if too acid it attenuates to a greater length. The sufficiently scalded curd is allowed to pitch for a quarter of an hour, and then a rack is put on and weighted with a 56-lb. weight. Thus the curd remains until it is consolidated or begins to mat. It is then cut up the centre with a long knife, rolled to the upper end of the vat, and the racks and weights placed on as before. Draw off the whey, remove the weights from the curd, cut it up and spread it on the bottom of the vat.

PACKING AND "CHEDDARING."—Replace the curd in a square block in the bottom of the vat, sweep up all the crumbs, re-weight and allow to remain ten minutes. Cut into bricks and remove to the curd-sink; cover with dry cloths and put on the weights. Open and turn every twenty minutes, turning the outside of the curd within. When the curd is firm and tough, cut it into two-inch cubes, tie up in a cloth, cover with dry cloths and a tin pan and apply the weights. Open out and separate every half-hour, using dry cloths each time until it is ready to grind. The above method of manufacture results in a more open and meaty cheese than that obtained by adopting the modern or Canadian plan.

CANADIAN METHOD.—In this method the whey is drawn off before any matting or consolidating takes place, and the loose curd is removed from the vat to a curd-cooler, where it is stirred until it is dry enough to mat, which, however, is a point rather difficult for inexperienced persons to decide. Matting goes on until the curd is ready to grind. A curd is ready to grind when it is distinctly acid to the taste and smell, dry and solid in cutting, tears stringy, and attenuates from 1 in. to $1\frac{1}{4}$ in. on the hot iron.

GRINDING is done to reduce the curd to such a condition that salt can be thoroughly distributed; it also allows of the cooling of the curd. When ground the curd is ready for weighing, and, if cool enough, for salting.

SALTING.—About two per cent. of salt is the amount usually added, and the temperature of the curd should not be above 80°. The salt hardens the curd, helps to dry it, has a slight antiseptic action and therefore arrests decay to some extent, and also has a tendency to check further development of acidity in the curd. After adding the salt stir the mixture well for fifteen minutes, which will ensure thorough incorporation of the salt and the curd. When the temperature of the curd is 70° to 75° it is ready for putting into hoops which are lined with a cloth. In filling the hoops press carefully with the closed hand. After the hoop is full place it in the press.

PRESSING.—The pressure must be gradually applied, and should reach 10 cwt. in two hours' time, at which pressure it is allowed to remain over night. If pressing is excessive during the first few hours, fat is expelled with the whey, and the quality of the cheese is lowered. Besides this, a hard firm coat round the external portion of the cheese is got, which checks the drainage of the whey. The object of pressing is to bind and consolidate the curd, and to expel whey. A suitable temperature in the press-room (60°) aids the objects of pressing. The morning next after the day of making, the cheese is taken out of the press, the cloth is removed, and the cheese bathed for one minute in water heated to 120°. This improves the condition of the coat, rendering it tougher and less liable to crack. After bathing put on clean cloths, and return to the press. Apply 10 cwt. pressure during the first two hours, and then 15 cwt. until next morning. On the morning of the third day turn the cheese, grease it, cap one end, and return to press with a smooth cloth; then apply 1 to 1¼ tons of pressure. The grease is applied to fill up cracks, to render the outside of the cheese smooth, and to enable the bandages to stick. On the fourth day turn the cheese, put a cap on the bare end, place in a clean cloth, and then apply pressure until the afternoon. In the afternoon bandage with a laced or winding bandage, weigh, and take up to the curing-room.

CURING OR RIPENING.—The temperature of the curing-room should be 65° to 70°. New or young cheeses require the higher,—old cheeses the lower temperature. The ripening-room requires to be kept at an even and correct temperature, for although the making of a Cheddar depends so largely on success in the first stages of the operation, there is yet a possibility of spoiling the best of curds if due attention is not given to the temperature of the ripening-room. When the temperature is too low the result is a soapy cheese lacking body and flavour; when too high, sweating occurs, loss of fat, and dryness in the cheese. The cheese must be turned daily for six weeks. Neglect to turn results in redness on the ends of the cheese, and moisture descends to the end which is resting on the racks. A certain amount of ventilation is necessary, but there must be no draughts. Usually the room is kept dark, which, however, is of little if any advantage, except that cheese-flies are not then quite so numerous.

CHAPTER VII

ON THE BEST METHOD OF MANUFACTURING STILTON CHEESE

THE process of making a Stilton cheese has more similarity to that of the manufacture of some of the Continental cheeses than any other British make. Despite this fact it is a British cheese, and the county of Leicestershire can justly claim the honour of being its home. Indeed many people consider that it is impossible to make the real article outside the county named. This, however, is an error, as with suitable buildings and utensils, with perfect cleanliness and with sufficient skill on the part of the maker, prime Stilton can be made in any district. The cost of producing a Stilton is however rather greater than that of a Cheddar or Cheshire. This is owing to the greater cost of the buildings, the greater amount of labour, the longer time taken in curing, and lastly, to the fact that less ripe cheese is obtained from a given amount of milk by the Stilton method than by the methods just mentioned.

The Stilton is popularly considered to be a cream-cheese, but at the present time it is nearly always made of whole milk without the addition of cream, and yet the quality produced leaves nothing to be desired. Nevertheless the milk intended for making Stilton should be of at least average quality, and that produced by cows grazing on rich old pastures is the most suitable. The giving of large quantities of cake to the cows is not to be recommended, as this usually produces a milk that causes trouble during the making of the cheese.

In the method of manufacture about to be described, two separately made curds are used. This method is the one by which the best Stiltons are made. One reason why this is so is found in the fact that separately made curds do not unite as closely as curds made at one operation. The consequence is that we get a great amount of air space in the body of the cheese, and therefore fulfilment of one of the conditions essential to the development of the mould which it is the pride of the Stilton maker to obtain.

Before commencing operations the maker should have in remembrance the leading characteristics of an ideal Stilton. These are as follows—A drab-coloured rough wrinkled skin, a texture salvy and mellow but not soapy (indeed, as the old Stilton maker's maxim says, "beware of chalk and beware of soap," which implies medium texture, and avoidance of hardness on the one hand and soapiness on the other), a marbling throughout the body of the cheese due to the growth of a blue mould (*Penicillium glaucum*), and the possession of an unique flavour.

The following is a list of the requisites for the manufacture of Stilton—(*a*) Building. The building or dairy must be divided into at least three separate apartments, or better still if into four. These are—(1) A setting-room and draining-room. One room may be made to serve the double purpose of setting and draining, or a separate room may be used for each

METHOD OF MANUFACTURING STILTON

purpose. (2) A drying- or coating-room. (3) A storing- or curing-room. Besides these a cellar is a great advantage, as the cheeses can be taken there when they are ripe, or even before they are ripe if the weather is hot, and the ordinary rooms are out of condition. For Stilton-making it is imperative that all the rooms should be high and well ventilated, and that they should be so constructed as to allow of cooling them in very hot weather. Further, they must have apparatus for heating purposes, as during spring and autumn artificial heat is a necessity. (*b*) Utensils. Briefly enumerated these are—(1) A renneting-vat made of tin; (2) a curd-ladle or scoop of about half a gallon capacity; (3) straining-cloths; (4) a curd-sink made of glazed earthenware; (5) a draining-sink lined with tin; (6) perforated metal moulds or hoops; (7) boards (9 in. × 9 in.); (8) draining-shelves; (9) turning- and bandaging-table; (10) knife, bandages, &c.

MANUFACTURE.—Milk. (The milk for Stilton-making should be perfectly fresh, and not slightly acid. as is the case in the making of some British cheeses. This necessitates the renneting of the milk as soon as received into the dairy, and that which has never lost its animal heat is the most suitable.

RENNETING.—The rennet is added when the temperature of the milk has fallen to 84° F.; and the amount required is $1\frac{1}{2}$ drachms to every 60 lbs. of milk. Most makers consider that prepared rennets are inferior to the home-made ones. Yet we know that the use of home-made rennets is not essential to the making of the best Stiltons, as these are constantly made from prepared rennets. It seems probable that in using prepared rennets the makers accustomed to the home-made article make no allowance for the greater strength of the former, and consequently add too much. This results in an inferior cheese, but the fault is due to the maker and not to the rennet. After adding the rennet to the milk, thorough mixing of the two should be brought about by stirring. Let this be continued ten minutes, by which time mixing will be complete and there will be no danger of any cream rising. Now allow the contents of the vat to set for 1 to $1\frac{1}{4}$ hours, according to the state of the curd. This, although a somewhat prolonged coagulation, is not unusual in the making of sweet curd cheeses.

CURD DRAINING AND DEVELOPMENT OF ACIDITY.—When ready, the curd is ladled out of the vat into straining-cloths, placed in the curd-sink. These cloths are about a yard square, and hold from three to four gallons each. In the act of ladling the curd is cut into thin slices, whereby the drainage of the whey is facilitated. The curd is allowed to stand for half-an-hour in its own whey, or longer if it is soft. The whey is then let off, and the curd tied up by bringing together the three corners of the straining-cloth and using the fourth as a binder; and here in the curd-sink it drains until evening. To aid the draining, tighten the cloths every hour during the first eight hours. This tightening requires to be done with care, so that no curd is crushed in the operation. In the evening the curd is cut up into squares of about four inches and laid in the draining-sink with a light cotton cloth thrown over it. Here it remains over night, and during this time it slowly oxidizes. The evening's milk is treated in the same manner as the morning's milk, being allowed to drain during the night whilst in the curd-sink. In the morning cut up the evening's curd, and then allow the two curds to develop the requisite amount of acidity. If acidity does not develop rapidly enough, tear up the curds to aid it, or place them upon racks and keep them warm with hot water.

SALTING.—When the curds are ready, *i.e.* when they have developed a sufficient amount of acidity, and are of a certain mellowness, they are broken up by hand into coarse-grained pieces. It is always difficult to decide when the curds are ready, and experience is the only teacher. The following, however, are some of the signs that guide the maker as to the fitness of the curds—The first curd made should be clean, flaky, decidedly acid, and free from sliminess or sponginess; the second should be in about the same condition, but not so acid. It takes usually thirty-six and twenty-four hours respectively before the curds show the above signs. After these are broken they are mixed together, and a rather coarse salt is added at the rate of about $1\frac{1}{2}$ per cent. by weight of the curd. If the curd is wet add more salt, if dry add less. It is usual to obtain 18 lbs. of curd from 12 gallons of milk.

HOOPING.—The curd, after a thorough mixing with the salt, is put into hoops holding 20 to 24 lbs. each. If the cheese is for sale in a wholesale market let it be made full-sized, as such cheeses are easier to sell than small ones. The temperature of the curd at the time of hooping should be about 60° F. Before commencing to fill the hoops, place them on a board covered with a piece of calico. In filling, the curd should be firmly pressed at the bottom, and lightly at the sides, and the larger pieces should be put into the loosely-filled centre. By taking these precautions a cheese is obtained that presents a good surface.

CHEESE-DRAINING.—When the hoops are filled, they are carried, together with the board and cloth on which they stand, to the draining-shelves. The temperature of the room in which the shelves are placed should be 65° F. The hoop and cheese should be turned after standing two hours, an operation performed by inverting them upon a board and cloth similar to those on which they stand. The turning should be repeated before leaving for the day, and it must be performed at least once each day for the next nine days. Neglect in turning at this stage causes unequal ripening of the cheese, and the ends become uneven. If the curd does not settle properly it should be skewered through the perforations in the hoop, and a little salt should be rubbed on each end.

SCRAPING AND BANDAGING.—In about nine days the cheese is taken out of the hoop, and if ready it is scraped with a knife. It is known to be ready for scraping when the cheese leaves the side of the hoop, when it is creamy on the outside, and when it has a smell similar to that of a ripe pear. The scraping makes a smooth even surface, fills up cracks, and aids in the production of the much-desired wrinkling of the coat of the cheese. This last result is brought about by the consolidating effect of the scraping on the surface of the cheese, and the comparatively loose and free state in which the central portion remains. In consequence of this difference the external portion of the cheese settles less

METHOD OF MANUFACTURING STILTON 89

than the internal portion, and consequently a wrinkling of the coat of the cheese follows. After the cheese has been scraped, a bandage is tightly pinned round it, a cap placed on its upper end, and the cheese is put back into the hoop. Next day remove the hoop and bandage, and scrape the cheese, then tightly pin on a clean bandage round the top. Allow the bandage to hang loosely down, invert the cheese, and loosely fold the bandage over it. The cheese is then put upon the draining-shelves without the hoop, and there it remains until the coat begins to appear, which usually happens about the eleventh day counting from the day of hooping.

FORMATION OF THE COAT. — About the eleventh day the external surface begins to show signs of white mould, also dry patches appear on the bandage. These are the first signs of the coat, and on their appearance the cheese is ready to go to the drying- or coating-room. This room should be cool and damp, have a temperature of from 55° to 60°, and if possible it should have a gentle, cool, moist draught passing through it. By thus keeping the air of the coating-room cooler and moister than that of the draining-room we minimize the loss of moisture, and consequently avoid lowering the quality of the cheese, and at the same time we prevent fermentation becoming too rapid. If the coating-room is too dry, and the cheese shows signs of becoming hard, cover it with a moist cloth. The cheese on going to the coating-room has no bandages on it, but there is the small cloth on the board on which it rests, and this requires changing each day when the cheese itself is turned. Turning goes on for a fortnight, and by the end of that time the coat should be firmly fixed.

CURING.—When the coat is firmly fixed, the cheese is ready to go to the storing- or curing-room, which may be an airy cellar, or a cool upper room kept at a temperature of from 55° to 60° F. If the temperature is too high excessive evaporation ensues, and as a consequence a hard dry cheese; if too low the ripening of the cheese is retarded. The shelves of the curing-room must be kept quite clean and free from mites, and the cheese turned daily. It takes a Stilton from four to six months to ripen, but some people try to shorten the period by skewering. This, however, is a rather doubtful proceeding, and yet it is permissible if the cheese is close, and there is a lack of mould-growth. When such a plan is followed, care must be taken that the apertures made in the cheese are closed up, so that flies and mites will not be able to enter. The skewers should be put into the cheeses from each end, not at the sides, and their ends should pass each other.

Besides this two-curd system of Stilton-making there is a "wet-curd" system. The essential difference between the two is to be found in the length of time during which the curd is allowed to stand in its own whey. In the wet-curd system the whole of the whey is not drained off until the curd is ready for vatting; whereas in the method just described the curd stands in its own whey about half-an-hour.

Before concluding, we may with advantage briefly sum up the points of difference in the making of a Stilton, and in that of the better known and much more widely made Cheddar. In Stilton-making the rennet is added to a perfectly fresh milk, in Cheddar-making to slightly acid milk; also less rennet is added if Stilton is to be made. It is owing to these two factors that the coagulation in Stilton-making is more prolonged than in the case of Cheddar. Again, in Stilton-making the development of acidity is not pushed by scalding as is the case with Cheddar, and instead of taking eight hours, it takes usually twenty-four and thirty-six hours. It may, however, be noted that in Cheddar-making acidity is allowed to develop in both milk and curd, whereas in Stilton-making it is only allowed to develop in the curd. Less salt is added to the curd of a Stilton than to that of a Cheddar, but this is more apparent than real, for when the curd of a Stilton is ready to salt it is much moister than that of a Cheddar. Lastly, the curd in Stilton-making is put to drain in a much softer condition than in Cheddar-making, but no pressure is applied to the former, whereas a ton and upwards is required for the latter.

Finally, we feel fully justified in stating that a

well-made Stilton stands without rival amongst the better known varieties of cheeses; and we know from experience that by the system just detailed it is possible to produce an article of prime quality.

CHAPTER VIII

CHESHIRE CHEESE-MAKING

CHESHIRE cheese is of more local than cosmopolitan repute; indeed the making of it is practically confined to Cheshire and the counties that border upon it. The locality in which this cheese is made is really restricted to that wherein a demand for it exists, as its fragile nature renders it unsuitable for exportation purposes. The general conditions as to the food of the cow producing the milk intended for Cheshire-making are similar to those applicable to Cheddar. The dairy required is also similar. It consists of three apartments—a making-room, a press-room, and a curing-room. The press-room in Cheshire-making, however, must contain, in addition to the presses, an oven, wherein the cheeses can be placed immediately after hooping. This so-called "oven" is merely a recess in the press-room wall, so situated as to have the kitchen fire at the back of it. The utensils required are such as are used in any process of cheese-making, but the hoops are usually perforated, the vat is jacketed and rather shallow, and the curd-mill is fine-toothed, so that the curd can be ground down to a rather fine state of division.

There are three methods of manufacturing Cheshire cheese, each of which produces a special type of cheese. The three methods are the early ripening, the medium ripening, and the late ripening, named after the predominant characteristic of the cheese produced, *i.e.* an early ripening cheese, a medium ripening cheese, and a late ripening cheese. The two latter of these cheeses are of much higher quality than the first named. Yet at the present time the quick or early ripening cheese is much made, and this probably is one of the causes of the prevailing low prices.

The method of manufacture about to be detailed refers to a cheese which will take about three months to ripen, and is therefore classed as a medium ripening cheese. The qualities looked for in such a cheese are—a rather high colour produced by the addition of colouring, a looseness, granulation, and openness in the body and texture known as "meatiness," a certain amount of crumbliness, and a mellow, rich, tasty flavour.

PREPARATORY TREATMENT OF THE MILK.—Strain the evening's milk into the vat, and keep it at such a temperature that it will be about 68° F. in the morning. In the morning skim off the cream, and heat it to 95° F.; then pour it along with the morning's milk into the vat. If the correct amount of ripeness has been developed (and this is of the utmost importance), rennet the milk; but if not, either keep the milk in the vat at a temperature of 94° F. until it is ripe enough, or add sour whey, which latter is the more common method. A little before this stage is reached the colouring is added, indeed it should be added ten minutes before renneting. When the colouring is added immediately before the rennet, there is great liability of getting a discoloured cheese. This, although one of the causes of discolouration, is not the chief one. At the present time white or uncoloured cheeses are being made, a method that is to be recommended, as it avoids all danger of discolouration from improper mixing of the colouring. But unfortunately the public demand is for a high-coloured cheese, and therefore colouring is still added by most makers, although it is so risky.

RENNETING.—When the milk is ready to rennet, it should give a rennet test of twenty-two seconds, which is rather longer than is required in Cheddar-making, or in other words the milk is sweeter. The temperature of the milk at the time of renneting should be 86° to 88° F., and the amount of rennet required is one oz. of rennet extract to twenty gallons of milk, or such an amount as will produce a curd that is ready to cut forty-five to sixty minutes from the time of adding it. After adding the rennet, stir the mixture in the vat for five minutes. Next cover the vat with a cloth, and when the curd is firm enough, cut it with the American horizontal knife, and then with the vertical knife until it is in a rather coarse condition. Just after the curd is cut, a little whey is usually drawn off for adding to the next day's milk, to aid the development of acidity. Next clean down the sides and bottom of the vat, and with the hands stir well for about fifteen minutes, when scalding should begin.

SCALDING.—The scalding is only partial, and the curd at the termination of it is considerably softer than that produced in Cheddar scalding. The scalding should be gradual, and a rate of 1° in five minutes is very suitable. Scald until a temperature of 92° to 94° is reached, and during the whole process careful and continuous stirring is required. When the correct temperature is reached, continue the stirring until the curd is quite firm, and the corners are rounded. Then allow the curd to settle for about an hour, or until it leaves the sides of the vat. Next cut the curd up the middle, roll it up to one end of the vat, and let off the whey.

DRAINING THE CURD.—When the whey has drained off cut the curd into blocks, and place it at one end of the vat. Then put a rack in the bottom of the vat, spread a cloth on it, and place the curd upon it, carefully covering it with a dry cloth. Turn the curd every ten minutes, and at each turning break it up into pieces of about two inches diameter. The turning is repeated until the curd is sufficiently dry and acid, and four or five turnings are usually required. After the last turning grind the curd twice, making it finer than the curd of a Cheddar. The object of the fine grinding is to produce the granular, open, crumbly texture that is so much sought after in a Cheshire.

SALTING.—Salt is added at the rate of 7 to 8 ozs. per 20 lbs. of curd, more being used if the curd is wet, less if it is dry. The temperature at the time of salting should be above 70° F., and below 80° F. If below 70° the curd will not take the salt, and the cheese will afterwards become black in the centre. If 80° or above, there will be loss of fat during the after treatment of the cheese. Thoroughly mix the salt and the curd, and then put the salted curd into a hoop lined with a coarse cloth. After hooping take the cheese into the press-room, and place it in the cheese-oven, where a temperature of 75° to 80° is maintained. Here the whey slowly drains from the curd, the curd itself contracts, and the amount of acidity gradually increases. The escape of the whey is facilitated by the insertion of skewers, and their occasional removal. After the cheese has been in the oven for four hours it is turned, put into a dry coarse cloth, placed back again into the oven, and there it remains until morning.

PRESSING.—Next morning the cheese is removed from the oven and put into a fresh cloth. It is then placed in a press, but no pressure is applied, or only a very little. On the next three or four mornings the cheese-cloth is changed, and the pressure is gradually increased. By about the fourth morning whey will have ceased to exude, and when such is the case the cheese should be removed from the press and taken into the curing-room.

CURING.—Before taking the pressed cheese to the curing-room a bandage is pasted on to it, the paste used consisting of flour, boiling water, and borax. Over this bandage, an ordinary cheese-bandage is placed, and the corners of the cheese are often ironed with a hot iron to render them smooth. When the cheese is bandaged take it to the curing-room, which should have a temperature of 60° to 65° F. The shelves of this room are frequently covered with straw, upon which the cheeses are placed. Such a plan tends to produce a growth of green mould on the external surface of the cheese. The cheeses require to be turned daily for the first week or two; then gradually lessen the number of turnings until once per week is reached, and this must be continued until the cheese is sold. The cheese will be ripe in about four months.

Although we have just detailed a method of making Cheshire cheese, no exact data can be really considered to represent the Cheshire method, since it varies considerably with different makers, and according to which one of the three kinds of cheeses it is intended to produce. The aim throughout each system is undoubtedly to produce the best article of its kind, and the following principles, considered along with the details of the medium process just described, will roughly indicate the variations that have to be made in order to produce the different types of Cheshire cheeses:—The milk to be moderately sweet when the rennet is added; the temperatures throughout the process of manufacture to be varied according to the moistness of the curd required; if a dry curd is wanted the temperature should be comparatively high, if a wet one comparatively low; the quantity of rennet used to be varied according to the time the cheese is intended to ripen in, more being used if for quick ripening, and less if for slow ripening; the size to which the curd is to be cut depends on the amount of whey that is to be left in it, and this again depends on the kind of cheese; for a quick-ripening cheese leave a deal of whey in the curd, and cut into large-sized pieces; for a slow-ripening cheese expel the whey thoroughly, and cut the curd into very small pieces; the amount of acidity to allow to develop in the curd whilst in the whey must be greater the sooner the cheese is required to be ripe; the size of the particles of curd on salting also to be varied according to the time in which the cheese is wanted to ripen; the shorter the ripening period the coarser should be the curd, and *vice versâ;* the pressure to be regulated according to the amount of whey required to be expelled, *i.e.* according to the dryness the curd is wanted; for quick ripening expel little whey, and therefore apply little pressure; for slow ripening expel all the whey possible, and therefore apply much pressure; a quick-ripening curd on hooping should be coarse-grained, and saturated with acid whey; a slow-ripening curd on hooping should be fine-grained, dry, and contain very little free whey.

Throughout the Cheshire systems the endeavour is to develop more or less of acidity after the curd is hooped, and hence the use of the oven.

The already described medium ripening process produces a good Cheshire cheese of such quality that when ripe it will keep a few months, should the markets necessitate such a plan. This clearly indicates one of the advantages of adopting this process (and this remark is also applicable to the late process), for should the early ripening one be adopted, the produce must be sold as soon as ripe, or else be wasted, as it has no keeping properties. On the other hand, the quick-ripening process produces a greater weight of cheese than the other two processes, and it also gives quicker returns, but the quality of the cheese produced by it is not first-class, and the risks as above indicated are great.

CHAPTER IX

WENSLEYDALE CHEESE-MAKING

THE making of this cheese is practically confined to the beautiful dales that render the north-western portion of Yorkshire so picturesque. As the name implies, the chief locality in which it is made is Wensleydale, and here the cheese has been made for centuries. This dale is not only famed for its cheese, but also for its variety of sheep, the so-called "blue-faced Leicester" or Wensleydale.

A study of this method of cheese-making shows us that the fine pastures of the Yorkshire dales, chiefly on soils derived from limestone rocks, are especially adapted for producing a first-class cheese-making milk. Apart from this nothing special is needed in the way of food for the cow producing the milk used in the making of this cheese; also no special dairy accommodation is required, and no special utensils are employed. In the old-fashioned method, a large brass or copper pan, called a "cheese-kettle," was used in place of a cheese-vat, but the use of this is fast dying out.

The cheeses are made of two shapes, "flat" and "Stilton" shape. The former of these are suitable for making during spring and autumn, and also when the cheeses are intended for immediate consumption. When the cheeses are made of the "Stilton" shape, they are supposed to develop a greenish-blue mould just as a real Stilton, but with the flats this is not looked for. The Stilton-shaped Wensleydales are therefore classed as British blue mould cheeses. The period of ripening of Wensleydales varies according to the shape adopted, but this is only so owing to the differences in the curds used in making the respective shapes. The "flats" take only a short time to ripen, the "Stiltons" a longer time. Although we speak of "Stilton-shaped" Wensleydales it is rare to find them exactly resembling a Stilton in shape, as the cheese usually becomes much distorted after its removal from the hoop. Indeed, some makers consider that irregularity in shape is a sign of good quality. Nor is this without reason, for, in order to acquire the distinctive characters of a Wensleydale, the curd must be hooped when it is in a moist condition, and only a small amount of pressure must be applied to the cheese; and these two factors render a cheese liable to unshapeliness.

A good Stilton-shaped Wensleydale possesses the following characteristics—A smooth surface, frequently a distorted shape, a soft, yielding texture similar to a Stilton but tougher, a blue mould evenly distributed throughout the body of the cheese, and not running in veins as in the real Stilton, and a mellow, creamy, mouldy flavour.

In the past there was no fixed method of making the cheese, but now teaching is aiding to bring about a definite system, and also it is raising the average in regard to the quality of the cheeses produced. Some good cheeses were formerly made, but there were also many bad ones, and the average was decidedly lower than that of the present time.

The method of manufacture about to be detailed is the modern method, and although the utensils used are not such as most of the dalesmen possess, yet they would undoubtedly be able to get a greater uniformity in their produce by using such. More especially would this desirable result be brought about if they gave attention to the quantity of rennet recommended; to the temperature of coagulation, of scalding, of the curd on salting, of the curd on hooping, &c.; to the amount of acid; and finally to the method of salting. The adoption of such particulars avoids the haphazard results of the old style of making.

PREPARATORY TREATMENT OF THE MILK.—Allow the evening's milk to run into the cheese-vat, and cool it down to 60°. Stir the milk occasionally during the evening, which will help it to cool, and will also prevent the cream from rising. In the morning skim the cream off the evening's milk, and heat it to 90° F. Then pour the morning's milk, and the heated cream along with it, into the vat amongst the evening's milk, and raise the temperature of the mixed milks to 86°—88° F.

This method of treating the milk is applicable to cases where making is followed once a day, and only in very hot weather need the cheese be oftener made. If an excessive amount of acidity develops in the milk, the cheese will be dry and hard, and will never possess the true qualities of a Wensleydale.

RENNETING.—Given that the temperature of the milk is as stated, and that the milk itself is perfectly sweet, the rennet may be added. One drachm of rennet extract to 40 lbs. of milk will produce a firm coagulation in about an hour, and therefore is the right quantity to add. After the addition of the rennet stir the mixture for five minutes. When the curd is sufficiently firm break it into cubes of about half-an-inch square, using American knives for the purpose. This breaking or cutting takes about five minutes, and after it is performed the curd is allowed to settle for five minutes. After settling, the curd is stirred for about twenty minutes with a shovel-breaker, rake, or hand. The latter of these is preferred when a small quantity of milk is being handled. After the stirring allow the curd to settle for ten minutes.

PARTIAL SCALDING.—Sufficient whey is now drawn off, so that when heated it will raise the temperature of the contents of the vat to what it was previous to renneting; the whey taken off should not be heated to more than 130° F. After adding the heated whey, stir constantly for about half-an-hour, and then allow the curd about twenty minutes to settle. It is not always necessary to even partially scald in the making of Stilton-shaped Wensleydales; indeed in summer-time it is only requisite when the weather is damp and cold, or whenever the curd seems as if it would be long in getting dry and firm. When "flats" are made scalding is always requisite.

In case of not scalding the curd, the stirring is longer continued, and the curd is given a longer time to settle. The whey is let off when the curd is in the right condition. This, however, is not easily described, and experience is the only guide. One sign of sufficient scalding is that you have 16 to 18 lbs. of curd from 12 gallons of milk. If more the curd is too moist, if less it is too dry. The whey is usually drawn off one and a half to two hours from the time of cutting the curd.

DRAINING AND DEVELOPMENT OF ACIDITY.—After drawing off the whey, take the curd out of the vat, and place it in a straining-cloth. Put it on a draining-rack, open it out after the first half-hour, and cut it into pieces; continue to do this every hour until the curd is ready to grind. A board is also placed on the curd whilst on the draining-rack, and 7 to 28 lbs. weight is placed upon the board. The amount of pressure is regulated according to the weather, and the drainage of the whey. When the weather is cold, and the drainage is slow, apply more pressure to the curd, and *vice versâ*. The curd when ready to grind should be decidedly sour, fairly dry and flaky, but not hard. It should be weighed before grinding.

SALTING.—The ground curd is salted at the rate of 1 oz. of salt to 4 lbs. of curd. The curd preparatory to salting is either ground in a mill or broken by hand, but in either case it must not be made too fine. The effect of fine grinding is a tight cheese in which no mould will develop. The time elapsing between adding the rennet and salting the curd is from six to eight hours. In the old system of making, the direct application of salt to the curd was only practised with large cheeses, the rule being to place the pressed cheeses in a strong brine, and leave them there for three or four days. The objections to this method are—(1) The uncertainty as to the amount of brine the cheese actually absorbs, as owing to differences in the amount of acidity present in the curd, the cheeses rarely absorb similar quantities, and as a consequence there is great variation in the cheeses produced. (2) The brine frequently does not penetrate to the centre of the cheese, and as a consequence a portion of it remains unsalted. (3) Placing cheeses in a brine cools them down to a very low temperature, and this interferes with the curing.

When direct salting of the curd is practised, it is necessary to allow a greater development of acidity than when brining is practised. The necessity for this arises from the rapid check of acid development when hand salting is followed, and the slow check of it when brining is followed.

HOOPING.—After grinding and salting the curd, put it into perforated tin hoops or moulds, without bottoms or with movable ones only. Place the hoop on a board and cloth, and loosely fill in the curd. The curd required to fill a standard-sized Wensleydale hoop is that which can be obtained from 14 gallons of milk. The temperature of the curd on hooping should be 64° to 65°. This comparatively low temperature is required in order to encourage mould development. Usually the cheese is put into the hoop without a cheese-cloth, but if the weather is hot it is better to use one. The cheese after being hooped is placed on a slab in the cheese-making room. Two hours after filling, the hoop and cheese should be turned and the cheese put into a dry cheese-cloth. Before leaving for the night a 4-lb. weight and a board are usually placed on the top of the cheese. When the cheese is left all night without pressure, or only with such as indicated, the temperature of the room in which it is placed should not be less than 60°.

PRESSING.—Next morning turn the cheese, put it into a dry cloth, place it in a press, and apply 1½ cwt. pressure for about five hours. Then remove it from the press, and turn it into a smooth cloth; replace it in the press, and apply 3 to 5 cwt. pressure until night. Next morning take it out of the press, sew on a bandage, and remove the cheese to a cool, moist room, placing it on a stone shelf. Let the cheese remain here for seven to nine days, turning it daily.

CURING.—Take the cheese to the drying- or curing-room, kept at a temperature of about 60° F. Turn the cheese daily, and if the weather is hot turn it twice a day. During the first few days it is necessary to skewer the cheese to prevent excessive heating. After six weeks in the curing-room, the cheese should be unclothed, and if the blue mould is not developing, the cheeses must be skewered. The skewering must be done from the ends, and after the operation care must be taken to cover up the entrance to the skewer-holes, to prevent the passing in of flies, &c.

The Stilton-shaped Wensleydales are ripe in four to six months; the flats are ripe in about two months. In the making of flats the curd is usually scalded, and is made much drier than if for Stiltons. The curd for Stilton shapes should be moist, slightly acid, and rather coarse at the time of hooping, whereas that for "flats" should be drier, more acid, and finer.

Properly made Wensleydales are prime cheeses, and there seems to be quite a possibility of their supplanting a deal of Stiltons within the next few years. This is not only on account of their possessing all the good qualities of a genuine Stilton, but also on account of the greater yield of cheese from a given amount of milk by the Wensleydale process, as compared with the real Stilton process. Indeed up to within the last few years Wensleydale cheeses were little known outside the locality of their making, but now that they are becoming of much wider repute, the demand for them is steadily increasing.

CHAPTER X

THE MILK INDUSTRY

THE production of milk in Great Britain is, next to the production of meat, the most important branch of our agricultural industry. During the past ten years it has attained gigantic proportions, and the old system of retailing milk to which water was frequently added, drawn from cows the great majority of which were stalled and fed within the precincts of our large towns, has given place to an improved system under which pure milk is dispatched, after production on the farm, direct from the rural districts to the distributor. The result is that the most valuable of all foods has been placed before the people of every class at a price within their reach, and under conditions which render it purer and safer than was formerly the case. What the consumption of milk was twenty-five years ago it is impossible to say, but in spite of the enormous increase in production, it is believed, upon the basis of careful estimates, that the consumption per head of the population per day does not exceed a quarter of a pint. In America the milk industry has increased with still more rapid strides, and in the great States of New York and Massachusetts the consumption has been raised by leaps and bounds, until, *e.g.*, the *per capita* consumption in the city of Boston has reached 1·33 half-pints per day. It is a curious fact that the milk consumption of the oldest city of the New World should be so much greater—in fact, nearly three times as great—as the consumption per head of the inhabitants of the city of Manchester.

The fact that milk production has been more profitable to the farmer than most of the other branches of his industry, has, of late, induced numbers of tenants to keep dairy cows and produce milk for sale, or manufacture butter or cheese. Increased production has in this way increased competition, with the result that prices have fallen. Hitherto the prices of cheese and butter have been regulated by the imports from other countries, but until recently there have been no imports of milk except in a condensed form, so that the price of milk, as retailed from day to day, has been regulated by home competition. Now, however, winter milk and cream are dispatched from Holland and Scandinavia, and although the quantity sent us is small as compared with the quantity we consume, yet it is evident that if a hundred thousand gallons can be sent successfully, it is probable that the trade will rapidly increase, and should this be the case the prosperity of the dairy farmer will decline more rapidly than it has hitherto done. We cannot prevent the importation of food from abroad, but we can control the system under which that food is imported, in order, first, that there may be no unfairness in the competition between our own people and the farmers of other countries; and second, that the food imported shall be pure and wholesome. In regard to the first point there is no fairness. The charges for conveyance of foreign milk and cream are infinitely less than the charges for the conveyance of the same materials at home; and as regards the wholesome condition of the milk sent us we have no guarantee whatever. Contagious diseases abound on the Continent, and farm workmen who have suffered from these diseases may be employed as milkers before they are fit for the work, while the fact that many of the cattle are diseased is sufficient to show that there is real danger in the consumption of imported milk which has not been sterilized before it is delivered to the consumer. We have remarked that home competition is intensified. The result is that prices have fallen to a figure which is without precedent, which means that unless dairy farmers combine to protect their own interests, prices will fall still further until no margin of profit remains. Nor is a reduction of price brought about by the action of the consumer, who in our experience has never sounded one note of complaint in this direction. It is owing to competition between the various competitors in the milk trade, so many of whom have striven to retain the retail price of milk and to pay the farmer, as they were accustomed to pay, such a sum as will enable him to conduct his business with success.

THE MILK INDUSTRY

By combination further reduction might be successfully resisted, but so long as action is isolated, the necessities of the milk trade resulting from the keenness of competition will ensure a further fall.

The milk represented by the butter and cheese we import amounts to nearly 1,245,000,000 gallons, whereas the milk produced for consumption, assuming the cows in this country—85 per cent. of which are in milk—to yield 400 gallons per annum, is 1,400,000,000 gallons. In a calculation made for a paper read at the Imperial Institute in 1895, I estimated the milk produced and sold as milk and in the form of butter and cheese at 1,405,000,000 gallons, so that the estimated yield on the basis above mentioned closely approximates to the estimated quantity consumed in some form or other. In the first place, the anual consumption of raw milk is placed at 13 gallons per head of the population per annum. The milk used in the manufacture of butter is estimated at 2·8 gallons to the pound; while the milk utilized in the manufacture of cheese is estimated at one gallon to the pound. If we add to this the milk used for condensing, and deduct 25,000,000 gallons, which I estimate to be the quantity displaced by the adulteration of whole milk with separated milk, we get the total to which we have already referred. Now, it is evident upon the face of these figures that we must necessarily import both butter and cheese in order to provide for the requirements of our people; at the same time, we are also shown that we have an enormous market if we can only provide the material for it. That material we should largely provide if the conditions were equal, but the foreign producer is assisted by defective British legislation, and by the unfair action of the railway companies, who carry his produce to the disadvantage of the English producer. There is no doubt that the consumption of milk will immensely increase, and the more the people realize that it is the most wholesome as well as the cheapest food in the world, the more readily will they increase their daily consumption. If they are shown, as they should be, as often as possible, that while a large proportion of the solid matter of meat is absolutely indigestible, and that apart from this there is considerable waste as between the joint purchased and the joint consumed—every particle of the solid matter in milk is digestible in the highest degree—they will be able to appreciate the fact that the one food is not only more valuable in the sustenance of mankind, but infinitely cheaper, pound for pound.

The cost of production of milk depends upon various circumstances, the rent of the land and its quality, the cost of labour and the cost of food. It also depends, particularly in winter, upon the manner in which the food is selected and utilized. Generally speaking, dairy cows graze in summer, grass being occasionally supplemented—and this is extremely wise—by the addition of cotton cake, grains, or meal, whereas in winter a common ration is chaff, pulped roots, with cake, meal or grains mixed and given, after heating for some hours, at the rate of so many pounds per day. The cost of the production of milk, then, depends upon the cost of the production of the hay, straw, roots, or whatever is grown upon the farm, as well as upon the cost of the purchased foods. It follows, therefore, that in producing milk, one of the chief objects of the dairy farmer should be to grow heavy crops of those materials which are consumed by the cow, and of which hay, straw, and roots are the chief in winter, and grass in summer. To this question, however, we cannot devote any space. It is, nevertheless, clear that those who feed upon a principle which has been found to succeed in practice, obtain the best results. They recognize that the cow needs the necessary material to maintain the heat of her body, to provide for the waste of tissue which is perpetually going on, and for the manufacture of the solid materials which are present in milk, and in consequence they compose a ration which includes the necessary proportion of albuminoids, which they obtain by using such foods as cake, beans, peas, vetches, or various meals with liberality. There is no doubt that the cost of production plus the cost of conveyance and the supply of railway churns closely approximates to the summer price of milk, which is perhaps upon the average no higher than 6*d*. a gallon, a great deal being sold below this figure.

We have referred to the system of adulteration which is now so widespread, and which is increasing from month to month. The Centrifugal

Cream Separating Machine, excellent as it is, has become, in some hands, a medium for the distribution of adulterated milk. When skilfully used, separated milk can be mixed with whole milk to a large extent and sold to the consumer without any fear of detection and punishment, and the reason is obvious. There is no standard of quality, and so long as a sample of milk satisfies the requirements of the public analyst it usually passes muster. The analyst is generally liberal-minded and generous, and in the absence of a definite law he frequently permits individuals to escape who ought to be severely punished. An average sample of good milk, from whatever part of England it may be taken, contains at least 3·4 per cent. of fat and 12·3 per cent. of total solids, but so long as a sample contains 2·75 per cent. of fat, and is not otherwise suspicious, it generally passes muster. There is practically no milk the produce of a well-fed herd of cows which contains at any time less than 3 per cent. of fat, but there are thousands of herds in which the average milk contains from 3·5 to 5 per cent. It is perfectly easy, therefore, to obtain milk of good quality, and by mixing it with separated milk to produce a mixture which, if analyzed, will be found to contain more than 2·75 per cent. of fat.

It has been urged by various responsible bodies that a standard requiring milk to contain 3 per cent. of fat should be fixed by law. I have myself urged that it should be raised to 3·25 per cent., and for reasons which are easily given. The trade insists that farmers as well as members of their own body would be frequently fined in consequence of the fact that milk occasionally falls below the proposed standard. It sometimes does fall below that figure in the case of individual cows, but a just law would provide that the owner of a single cow should be allowed to appeal to her, a sample of her milk being taken direct by the analyst, or in his presence. As regards the farmer, however, the matter is entirely different. It is in his power to select his cattle, to dispose of producers of poor milk, and to replace them with producers of rich milk, which are common enough; but no such steps are taken by the farming community to-day simply because the law does not control the quality of milk, and so long as anything will suffice which is passable, farmers cannot be expected to take trouble which will not increase their receipts. By the aid of recent inventions, farmers and milk-sellers alike are able to test a number of samples of milk in a few minutes, so that there would be no excuse for the distribution of a sample containing less fat than the standard required. If the present system is allowed to continue, the whole milk trade will degenerate into more or less fraudulent competition connived at by the authorities in power. If there were no precedent for the proposal which has been made, it would be more difficult to urge its expediency, but standards exist in many parts of the world—in America in particular; and it is remarkable that in Boston, Massachusetts, where the consumption of milk is greater than in any part of England, the standard is higher than in any other city in the world. After myself investigating the question in America, and being shown by those responsible for the conduct of the law that the consumption of milk as well as its quality has immensely increased since the institution of a high standard, I am satisfied that, bearing the above suggestions in mind, we should benefit the dairy industry of England in a high degree by instituting a standard for ourselves.

CHAPTER XI

THE PRINCIPLES OF BUTTER-MAKING

A SAMPLE of pure butter should contain no more than from 10 to 15 per cent. of moisture, a good sample averaging about 12 per cent., and, unless heavily salted, an almost infinitesimal proportion of mineral matter. Theoretically, butter should contain nothing more than the fat of milk, the salt which is added during manufacture, and the water which up to a certain point is inseparable from butter. Those who understand the manufacture of butter are well aware that both by the exercise of skill and carelessness a much larger amount of water can be added to it than is essential; and it follows that the larger the amount of water, the greater the weight of the butter produced. To knowingly manufacture butter with excessive moisture is fraudulent, for the consumer pays the price of butter for water; but it should be remembered that the perpetrators of a fraud of this character often defeat their own object, inasmuch as butter of high quality cannot be produced, nor will it keep if the water is excessive. Excessive salting is equally deleterious to the quality; a minute proportion of salt improves the flavour common to butter, but a large quantity masks it, at the same time adding to the weight. We have remarked that there should be no other material in butter than fat, water, and salt. In practice, however, it is next to impossible to remove either the whole of the sugar, or the casein or curdy matter; and this being the case, in the course of time—and it depends entirely upon the proportion of caseous matter left in the butter—a sample becomes rancid and unfit either for sale or consumption. The prime object, therefore, under the British system of butter-making is to produce as large a quantity of butter of the finest flavour as possible, reducing the moisture and the extraneous curdy matter and sugar to the lowest possible proportions. In the first place, then, in order to produce quantity it is necessary to use the cream separator, which extracts more fat from the milk than is obtainable by any other process. If this is followed by treatment which has for its object the conversion of as much of this fat as possible into butter, a maximum quantity will be obtained. As regards quality, it is first of all important that the milk should be obtained from carefully fed cows which are milked by clean hands into clean vessels, the milk being subsequently strained before manipulation. The apartment in which the various operations take place should be perfectly pure. In this case the cream from the separator will in due course ripen properly, and fine flavour will in consequence develop. Having obtained quantity and flavour, we have next to deal with the conversion of the butter-fat obtained in the churn into made-up butter. As we shall see, the grains of fat as they are first produced are floating in buttermilk, the particular constituent of which is casein. This casein is an essential food of the lactic ferment; hence its removal is necessary. Careful washing, therefore, is the first process; and if the tiny grains are washed at a given stage, which is shown in every dairy school, the greater portion of the curd will be removed, and almost pure butter-fat left behind.

Let us, however, assume that inferior butter is produced in a dairy, and that the occupier is unable to improve the quality. It may be asked how the production of an inferior article can be converted into the production of one of really high quality. The thing is easy if the work is carried out with intelligence and thoroughness. The manufacturer must condescend to details and recognize scientific facts. The alteration which takes place in cream, that is to say its change from perfect sweetness to a condition of sourness, acidity, or ripeness, is owing to the presence of an organism or bacterium which can only be discovered by those who are skilled in the use of the microscope. This organism rapidly increases in number when milk is warm and exposed to the atmosphere. It converts the sugar of milk into lactic acid; hence the sourness of milk. If this change is allowed to continue unchecked, the curd of the milk will coagulate, and it is for this reason that cream when allowed to ripen for churning becomes thicker. If cream is churned while it is still sweet it is frequently longer before it is converted into butter, it produces less butter, and the flavour is less full and nutty. The object, therefore, of ripening cream is to increase the quantity of butter and improve the flavour. In every dairy the lactic ferment is present either upon the utensils or in the atmosphere itself; but in some cases there are other organisms which, unlike the lactic ferment, have a contrary influence, producing a disagreeable flavour which reduces the value of the butter. The object of the dairyman, therefore, should be to maintain the apartment in which the milk or cream is placed, as well as the utensils employed, in as cleanly a condition as possible. There need be no fear about boiling water or lime destroying the lactic ferment. If it is removed from the utensils it is present in the air, and present, too, in a clean dairy perhaps in much larger numbers than any other organism is likely to be, and it is absolutely essential to the production of good butter. On the other hand, in a dirty apartment and on dirty utensils dangerous ferments are common; and if through conditions which suit them—and dirt is the chief of these—they are induced to increase in number, they may obtain the mastery, and destroy the flavour and quality of the butter produced. Let us suppose, therefore, as we have suggested already, that bad butter is produced in a dairy which has not been kept under the most perfect conditions. How can a change be brought about? In the first place, the whole of the utensils, shelves, and tables should be removed and thoroughly cleansed with boiling water. The walls and ceilings should be lime-washed and the floor scalded and dried, for a dairy should be dry. In this way every colony or nest, as it were, of the undesirable bacteria will be destroyed, and the clean utensils being returned to the dairy may be employed both in the raising of cream and in the manufacture of butter without any fear whatever. If, however, the manufacturer desires to proceed upon still more definite lines, and to omit no course of procedure which will ensure success, he may introduce from the most successful dairy with which he is acquainted a small quantity of the sour buttermilk which has been produced from the same day's churning. This buttermilk will contain the germs or bacteria which have been responsible for the production of butter-flavour of high class. If this buttermilk is added to the cream which has been obtained from the milk in the now thoroughly clean dairy, that cream will be inoculated, and when it has ripened it will be sufficiently perfect to be churned with every hope of success; and henceforth, so long as cleanliness is observed, there need be no fear as to the maintenance and constant reproduction of the friendly bacteria which are so desirable, as we have pointed out, in the manufacture of butter.

Let us now deal with the actual process of manufacture. The milk is drawn from the cows, and arrives in the dairy at a temperature of about 90° F. or a little higher. It may be at once passed through the mechanical separator and skimmed, or it may be poured while still warm into shallow vessels in order that the cream may rise by gravitation. Under such conditions the dairy should not be more than 60° F.—if it is as low as 50°, so much the better. The reason is that the greater the difference between the temperature of the milk and the temperature of the dairy the quicker and the more effectually will the cream rise. Cream is present in milk in the form of tiny globules; these globules are much lighter than the other portion of the milk, hence when the milk is at rest they rise to the surface just as a cork rises to the surface of a volume of water at the bottom of which it has been placed. The reason why the fat rises better in warm milk placed in a cold apartment is that the fat feels the change of temperature less rapidly than the rest of the milk, inasmuch as it is a non-conductor of heat. This being so, the difference in the density or specific gravity of the fat and the liquid portion of the milk is greater, and the fat is relatively lighter than it would otherwise be where there is no difference in the temperatures. In hot weather cream rises with far greater rapidity than in cold; milk rapidly becomes acid, both cream and milk thicken or coagulate, and for this reason the smaller globules of fat which are at the bottom of a milk-setting vessel are not able to rise at all—they are impeded, as it were, by the coagulation of the casein, hence a proportion of the butter-fat is lost to the churn. When, however, cream is raised upon a shallow vessel, it forms a thin layer on the surface and is brought into direct contact with the air, and consequently is oxidized or ripened with greater perfection: on the other hand, where cream is obtained through the medium of the separator it is kept in bulk and is less throughly oxidized, because in passing through the machine it has been in contact with the air for but a few seconds, while the air does not so thoroughly affect the mass of cream which is kept in a particular vessel as it does when the same cream is raised over a large area on the milk in a number of vessels. It is next to impossible to describe the exact flavour and appearance of cream which is just ripe for churning. Those who desire to know what it is like should take a lesson from an expert—and fortunately there are now plenty of teachers in almost every county in England.

When ripe the cream is passed through a strainer into the churn, and churned at a temperature which varies in accordance with the season of the year. In summer it may be churned at 56° F. and in winter as high as 64° F., but the exact degree depends upon the heat of the atmosphere, as we have suggested: a little experiment will enable the operator to thoroughly understand this point. Mixed cream should never be used: *i.e.* sweet and sour cream mixed together. The churn should be well cooled in summer and slightly warmed in winter by the aid of clean water, and let us remark that nothing is of greater importance than pure water; if it is impure, containing organic matter, this matter will be imported into the butter and will assist in decomposing it. After churning gently for a few minutes the carbonic acid gas which has formed in the churn may be allowed to escape by pressing the ventilator. Churning then continues until the grains of butter have reached the size of rice. At this point great care must be exercised. Some excellent makers here add a few quarts of very cold pure water, which gives crispness to the grains, preventing their adhering to each other so completely. The butter-milk is then drawn off through a sieve and more cold water added. It should be sufficient to enable the grains of butter to float in the churn and to partially harden. The water is then again drawn off and fresh cold water added two or three times, the churn being turned gently that the butter grains may be washed, although they should not unite and increase in size. Lastly, thin brine may be added, and in this the butter may remain for some little time before it is removed, or the floating butter may be removed from the brine with a scoop and placed upon the butter-table, or into the butter-drier or *délaiteuse*, from either of which the water is removed, by working in the one case, and by centrifugal force in the other. If dry salting is now performed the salt should be weighed, having previously been thoroughly rolled as fine as possible, dried in an oven and rolled again. It may be distributed by the aid of a dredger over the butter at the rate of half an ounce to the pound. If the butter is to be salted for keeping, from three-fourths of an ounce to an ounce may be used to the pound. The water having been perfectly expelled, the butter is made up for the market, or it may be allowed to remain in a wooden trough to still further drain, or it may, as in Denmark, be made up into rough rolls, allowed to harden for five or six hours, again worked, and finally made up for sale.

CHAPTER XII

CREAMERIES AND FACTORIES

A CREAMERY is generally understood to be an establishment in which the cream sent by the producer is converted into butter, whereas the entire milk of the farmer is handled in a factory, either for conversion into butter or cheese, or both, as may be found most convenient. The creamery system, which has been adopted in America on a somewhat large scale, has certain advantages which are worthy of notice, although when compared with the modern factory, these advantages are more than counterbalanced by the disadvantages. For example, the farmer who supplies a factory is required to deliver the milk twice daily. If he resides some miles from the building, it becomes essential to keep a horse and cart for the purpose, while the time of a man is very largely occupied on the road. The cream-supplier, however, is not required to deliver his produce daily. He removes the cream from the milk on his own farm and retains it, in accordance with the regulations, for perhaps a couple of days, taking care to mix every skimming carefully with the bulk. Thus the journeys are diminished in number, while the weight carried on each is incomparably less. In supplying the factory, too, the farmer either parts with the separated milk, which is a great loss to his stock, or he buys it back at a price which is often higher than it ought to be, while in all cases it has to be carried back to the farm. Formerly cream was purchased in America at so much per inch, but as cream differs in quality, this was found to be an unsatisfactory system. Latterly, the principle of churning the cream of each contributor separately has been adopted, with payment in accordance with the butter produced. I have had the advantage of inspecting creameries where this plan has been carried out, and of ascertaining from the books that not only was the quantity of butter produced very often exceptionally small, but its market value varied enormously, sometimes falling as low as

7*d.* per pound, and at other times reaching as much as 11*d.* The difference was owing almost entirely to the system prevailing on the various farms. Where care was taken to produce absolutely pure, clean milk, to raise the cream in an equally pure dairy, and to ripen it properly, the result was butter of high quality; but where no care was taken disagreeable flavours were developed in the cream, and the butter was in consequence immensely reduced in value. It must be evident that where cream of varying qualities are thus separately churned a dairy organization is placed at a great disadvantage. Success depends so much upon high quality all round and upon the acquisition of a name for a perfect sample. The very fact of a creamery turning out a variety of samples differing in quality, is sufficient to handicap it so seriously in the market that even the best butter it produces realizes less than would be the case if the whole of the produce were alike good. It may be safely pointed out, however, that although the produce of a butter factory is of much higher average quality than the samples of butter made in a creamery, that quality is to some extent controlled by the fact that the milk is mixed. For example, assuming that fifty farmers contribute milk to a factory, there are certain to be some who do not grasp the fact that the quality of butter depends almost entirely upon the purity of the atmosphere in which the cows exist, of the water they drink, of the vessels into which the milk is poured, and of the cleanliness of the udder, the hands and clothes of the milkers. The introduction of a few lots of unclean milk into a volume of clean milk will immensely depreciate its value and the value of the butter obtained from it. Thus it happens that the careless contributor obtains as much for his milk, if it passes muster, as the man who exercises the greatest care and skill. We may take it for granted that the creamery system is next to impossible in England, although it may still answer in some parts of Ireland, where, in consequence of the difficulties of locomotion, it is impossible to deliver milk twice daily. The same objection cannot be made to the factory system, which is the only system applicable to such countries as Canada, Western America, and the Australian Colonies, as well as to parts of Ireland and a few districts in England and Scotland.

Hitherto the reason why the factory has failed to obtain a hold upon the milk-producing portion of our population is that by means of butter or cheese production on a large scale it has not been possible to pay the producer so much for his milk as he can obtain by selling it for consumption in the large towns. There is, however, it is to be feared, a possibility that the value of milk for consumption will fall still lower until it approximates in value to the price paid to the factory. Should this be the case, the factory system is certain to extend; but under present conditions it is applicable—unless in a few special instances and for reasons which it is unnecessary to state—only to those parts of the country which are too distant from large centres of population, or which are badly supplied with railway communication. It is for this reason that we find factories existing in parts of Somerset, Devon, and Cornwall, and it is obvious that where butter realizes no more than from 10*d.* to 11*d.* a pound during several of the summer months, the milk cannot be worth anything like so much as its market price for consumption, which has seldom fallen below 6*d.* a gallon until last year, especially where it requires 2½ gallons to produce a pound of butter. Let us refer for a moment to what has been done in Ireland by the Irish Agricultural Organization Society. Last year there were fifty-six co-operative dairies with eight branches, while some fifteen other dairies were in course of formation. The returns obtained by the Secretary of the Society show that the average yield of the cows from which the factories obtained their milk is 435 gallons per annum. It has been pointed out that the average value of farm-made butter in Ireland in 1894 was 8*d.* a pound, so that, on the assumption that each pound of butter produced at home required three gallons of milk, the return per cow to the farmer would be £4 16*s.* 8*d.* On the other hand, the price paid for the milk by the factories having been 3½*d.* per gallon, the farmer contributing received £1 10*s.* 2*d.* per cow more than had he retained his milk for conversion into butter at home. In each case the separated milk would be utilized upon the farm, although we are bound to point out that something must be allowed for the conveyance of the milk between farm and factory. Estimating the value of the separated milk returned to the farm at 1*d.* per gallon, and the quantity returned at 345 gallons, the total receipt per cow would be £7 15*s.* 7*d.* If we take the average value of milk sold in England for consumption at 7*d.*, and it is possible that it will not reach a higher figure in the immediate future, we arrive at a total of £12 13*s.* 9*d.*, from which the cost of railway churns and of the conveyance of the milk to the station must be deducted. Again, a cow yielding the same quantity of milk would, if that milk was of exceptionally high quality, as in the case of the Jersey, produce 24 lbs. of butter, which, at the Irish price of 8*d.* per pound, would yield £8 0*s.* 8*d.*, or, plus 90 per cent. of the skimmed milk, £9 13*s.* 2*d.*; while at an average of 1*s.* a pound the return would equal £13 13*s.* 6*d.* How small, however, are all these figures compared with what was obtained a few years ago by butter-makers, cheese-makers, and milk-sellers alone! We have been enabled to examine accounts of dairy farmers in the county of Cheshire, where over £20 a head has formerly been realized on herds of from eighty to one hundred cows. As, in many cases, rents have not decreased, as labour has maintained its value, and as the reductions in food stuffs and manures have been comparatively small, it would seem that the deficiency is to be made up out of the farmer's pocket. The Irish factories realized an average price of 10·22$d.$ per pound for butter in 1894, which was 1·29$d.$ per pound less than in the previous year. While prices have fallen, however, the quality of the milk has risen. Thus, in 1893 the Irish factory milk produced 6·19 oz. of butter per gallon, but in 1894 6·33 ozs., and this is one feature to which closer attention will have to be paid in the future. Prices cannot fall below a certain figure; and it is possible that the farmer may, by the exercise of higher and still higher skill and care, not only increase the yield of milk per cow, but the quality of that milk also. It is, for example, quite possible to maintain a herd of Jerseys which will yield a still higher average than 435 gallons per annum, and at the same time to produce milk which will yield butter at the rate of one pound per $17\frac{1}{2}$ lbs. of milk, instead of one pound to $2\frac{1}{2}$ gallons, as is now the case in the best factories. In Ireland the cost of production based upon the general working expenses is about 10 per cent., or 1*d.* per pound when butter is at 10*d.*; but in the respective factories this figure may slightly differ for several reasons. Taking the factories working in 1894 as an example, it is found that the value of the butter produced varied from an average of 9·63*d.* to 11·60*d.* per pound, while the cost of the milk varied from 3*d.* to 3·83*d.*; and still further, the yield of the butter produced from the milk varied from 5·82 oz. to 6·78 ozs. to the gallon. Thus, then, the cost of production depends upon the quality of the milk, the price paid per gallon, and the market value of the butter, as well as upon the actual amount of the working expenses, such as wages, packages, machinery, wear and tear, carriage, and management.

We have already referred to the fact that cream has been paid for by the inch. It is now generally recognized that milk should be paid for in accordance with its butter value, or, where cheese is made, by its cheese value. Let us see how this system can be worked, taking the case of a factory paying a regular price to its contributors of 6*d*. an imperial gallon, or what is preferable, 6*d*. per 10 lbs., for the measurement of milk is never satisfactory, the fluid being larger in volume when it is warm and smaller when it is cold. Let us suppose, too, that the factory manager is willing to pay an extra penny per gallon to be distributed among the contributors in proportion to the value of their milk, no milk being received which does not contain more than 3 per cent of fat. During the three months over which the accounts run, 108,000 gallons have been delivered, representing an average of about 1200 gallons a day. The extra penny per gallon upon this quantity would amount to £450, which is the sum available for distribution among the various contributors upon the basis of quality. One contributor may have supplied 60 gallons a day with an average fat percentage of 3·2. Another may have supplied 30 gallons with an average percentage of fat of 3·5, while a third contributor has sent in 20 gallons daily, containing 4 per cent of fat. Now we shall see how the account stands—

	Gallons.	Days.	Per Cent.	Degrees.
A	60	90	3·2	1,728
B	30	90	3·5	945
C	20	90	4·0	720
Remaining contributors				34,300
				37,693

Each degree of fat is shown to be worth 2·86*d*., so that C, who has contributed 1 per cent of fat more than the minimum permitted, becomes entitled to £8 11*s*., multiplying the 720 by 2·86. He receives more than the full penny allotted, whereas, had his milk failed to reach more than 3 per cent. he would not have received a single shilling. This plan, which is ingenious and practical, is of American origin, and if we do not think our cousins equal us in the quality of their produce, they at least exhibit much greater skill and originality in their system of management. Let us take another system. Assuming that the whole of the milk delivered to the factory is of good quality, it is paid for in accordance with the butter actually produced. As it is impossible, unless the milk is separately handled, to ascertain the exact amount of butter produced by each of the contributors of milk, every sample is tested by such a machine as the Babcock Tester, and the quantity of fat present ascertained. Supposing A has produced 3000 pounds of milk containing 4 per cent. of fat, and B 4000 pounds containing $4\frac{1}{2}$ per cent., A will have produced 120 lbs. of fat and B 180 lbs., or 300 lbs. in all. It follows, therefore, that when the money value of the butter produced, after deducting the cost of manufacture, is distributed, B will receive exactly two-thirds as much as A, B's larger cheque being entirely owing to the fact that his milk was richer, and had produced more butter.

We are bound to remember that excellent as factory butter and cheese are, as compared with the average farm-house samples, there can be no question about the fact that the finest sample made by a skilled maker from milk which has been produced under his own supervision is superior to any sample produced in a factory which is necessarily obtained from mixed milk, produced under various systems of feeidng from cattle managed in different ways, and by more or less cleanly individuals. The factory is of enormous value, but we cannot admit that it can or that it ought to beat the produce of the farm, where that produce is obtained by the aid of the greatest skill.

THE END

Richard Clay & Sons, Limited, London & Bungay.

THISTLE MILKING MACHINE.

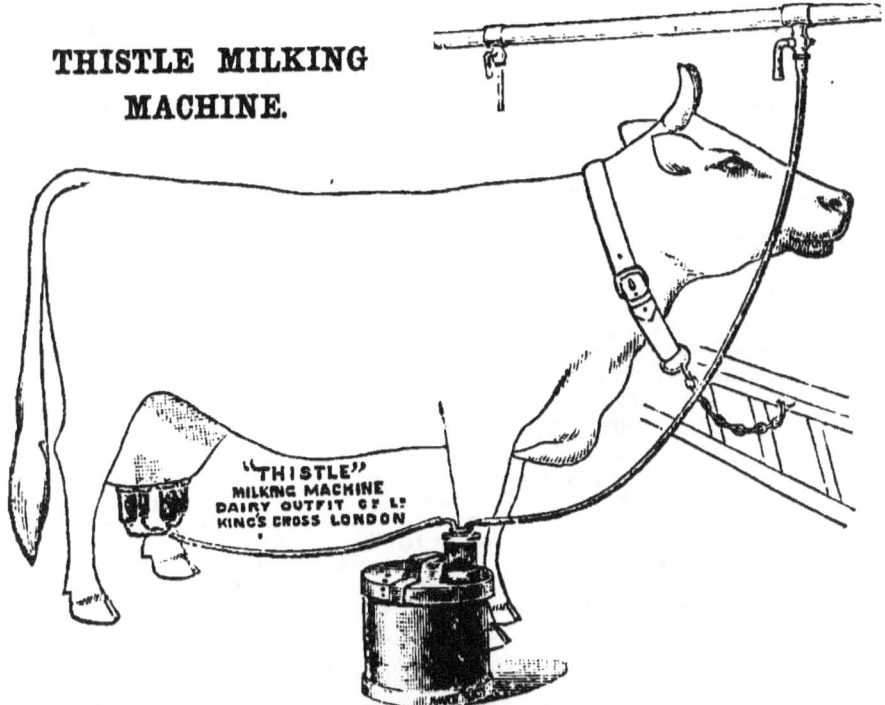

Sole Agents :—DAIRY OUTFIT CO., LTD., KING'S CROSS, LONDON,
Who also supply the **Victoria** and **Empress Cream Separators**
and all Appliances for Dairy use.

BRITISH DAIRY-FARMING,

TO WHICH IS ADDED

A Description of the Chief Continental Systems.

By JAMES LONG.

With numerous Illustrations. Crown 8vo, 9s.

"By far the most important part of Mr. Long's valuable contribution to the literature of dairy-farming is that mentioned in the sub-title of the book, 'A Description of the Chief Continental Systems.' By this comparison we do not intend to disparage the chapters relating to British dairy-farming, which are full of useful facts, figures, hints, and illustrated descriptions of most approved dairy implements and appliances; but a great deal of this is over old familiar ground, whereas in his chapters on Continental systems of dairying Mr. Long introduces us to fresh fields and pastures new. . . . Mr. Long has travelled in France, Switzerland, and Italy, with the special object of studying the manufacture of the cheeses for which these countries are famous all over the civilized world, and he has given such complete details in the book before us that it will be the fault of his agricultural readers if they do not make some of these fancy products of the dairy. . . . He has also a great deal to tell his readers about butter-making in France, Denmark, and other parts of Europe. His book is not a large one for his comprehensive subject; but it is crammed with valuable information which every dairy-farmer would do well to study."—*Pall Mall Gazette.*

FARMING FOR PLEASURE AND PROFIT.

By ARTHUR ROLAND. Edited by WILLIAM ABLETT.

Eight vols., large crown 8vo, 5s. each.

Dairy-Farming, Management of Cows, etc.
Poultry-Keeping.
Tree-Planting, for Ornamentation or Profit, suitable to every soil and situation.
Stock-Keeping and Cattle-Rearing.
The Drainage of Land, Irrigation, and Manures.
Root-Growing, Hops, etc.
Market-Garden Husbandry.
The Management of Grass Lands, Laying down Grass, Artificial Grasses, etc.

"This is another, and probably the last, of the series of agricultural handbooks, which are convenient in form, handy in price, and bring the information fairly up to date. The truthful illustrations of the various plants to be used, the preparation of the soil, the cultivation of the crop during its early stages, the means by which permanent fertility may be maintained—these are all matters which are clearly dealt with. The treatment of meadows, haymaking, &c., are very fully entered into, as also the cultivation of artificial grasses, fodder crops, &c."—*Field*.

Uniform with the above Series.

BRITISH BEE-FARMING: ITS PROFITS AND PLEASURES.

By JAMES F. ROBINSON.

With 21 Illustrations. Large crown 8vo, 5s.

"It puts before the public, on unquestionable authority, an unvarnished and honest statement of the advantages and difficulties (such as these are) of bee-keeping, and is in itself, therefore, a far more powerful incentive to the industry than any number of philanthropic or sentimental exhortations to the same object."—*Daily Telegraph*.

In small crown 8vo, 3s.

THE PLEASURES AND PROFITS OF OUR LITTLE POULTRY FARM.

By G. HILL.

"A charming picture of rural life."

"This is not by any means a dry collection of statistics, garnished with frequent tables bristling with figures. On the contrary, it is a very pleasantly written record of the successful experiments in poultry-farming made by a gentleman who had settled down on a small property in the north-east part of Hampshire. There is an abundance of useful information for those who are interested in the keeping of poultry."

CHAPMAN AND HALL, LIMITED.

www.ingramcontent.com/pod-product-compliance
Lightning Source LLC
Chambersburg PA
CBHW030316170426
43202CB00009B/1027